CLOVIS BLADE TECHNOLOGY

Texas Archaeology and Ethnohistory Series
Thomas R. Hester, *Editor*

CLOVIS

A Comparative Study of

BLADE

the Keven Davis Cache, Texas

TECHNOLOGY

by **Michael B. Collins**

with a chapter on microscopic
examination of the blades

by **Marvin Kay**

University of Texas Press
Austin

Requests for permission to reproduce material
from this work should be sent to
Permissions
University of Texas Press
Box 7819
Austin, TX 78713-7819.
utpress.utexas.edu/index.php/rp-form

LIBRARY OF CONGRESS CATALOGING-IN-PUBLICATION DATA

Collins, Michael B., 1941–
    Clovis Blade technology : a comparative study of the Keven Davis Cache, Texas /
by Michael B. Collins ; with a chapter on microscopic examination of the blades by
Marvin Kay.
        p.   cm.—(Texas archaeology and ethnohistory series)
    Includes bibliographical references and index.
    ISBN 978-0-292-71235-5
    1. Keven Davis Site (Tex.)   2. Clovis culture—Texas—Navarro County.   3. Clovis
points—Analysis.   4. Clovis points—Classification.   5. Excavations (Archaeology)—
Texas—Navarro County.   6. Navarro County (Tex.)—Antiquities.   I. Kay, Marvin.
II. Title.   III. Series.
E99.C832C65   1999
976.4'282—dc21   98-47835

Designed by Allen Griffith/EYE4DESIGN

ISBN 978-0-292-75799-8 (library e-book)
ISBN 978-0-292-78974-6 (individual e-book)

for GLEN EVANS *and*
TOM CAMPBELL—*teachers*

CONTENTS

FIGURES

TABLES

PREFACE

This study of Clovis blades from the Keven Davis and other sites has benefited from the contributions and assistance of a number of people. Obviously the primary debt of gratitude is to Keven Davis for bringing his original find to the attention of the archeological community, for sustaining a part of the subsequent investigations at the site, and for donating this important collection to the State of Texas. Bill Young, too, in his role as archeological steward for northeastern Texas, deserves special credit for recognizing the importance of Davis's find and taking a lead in documenting the site and the artifacts. Young also has provided much information on the site and the region in general. He and Davis have maintained abiding interests in this study and often shared information and offered encouragement. Young, Davis, and Bobbie Jean Young did the back-breaking excavation at the site.

Several colleagues are acknowledged for sharing information and insights during the analysis. These include Bruce Bradley, Robert J. Mallouf, Dee Ann Story, the late J. B. Sollberger, Thomas R. Hester, Glenn Goode, C. K. Chandler, Gene Titmus, Chris Lintz, Phil Wilke, Leslie Quintero, Dennis Stanford, John Broster, Reid Ferring, Ken Brown, Paul Tanner, and C. Britt Bousman. Paul Takac, Charles Hubbert, Al Redder, Scott Meeks, Boyce Driskell, C. K. Chandler, and Bruce Huckell shared unpublished data on blades and blade cores that add significantly to the comparative effort of this study.

The numerous illustrations by Pam Headrick and Frank A. Weir greatly enhance the descriptions attempted in the pages that follow, as do drawings by Karim Sadr (Fig. 6.7 a) and Ellen Atha (Figs. 3.3 b–d and 3.4 a). In addition to Pam Headrick, other staff members at the Texas Archeological Research Laboratory (TARL) have been extremely helpful. These are Laura Hillier Nightengale, who helped in locating and loaning for study artifacts in the TARL collections, as well as Milton Bell, whose photographs appear in Figs. 3.9, 3.10, and 3.11.

Additionally, Elizabeth Andrews printed many of the other photographs, and Edward Baker entered and processed the data used in Table 6.1. Marvin Kay prepared the plates used in Chapter 7.

Mallouf arranged for the Antiquities Permit under which this study was conducted, shared unpublished data, loaned for comparative study the Yellow Hawk collections and read an earlier draft of this report. That earlier draft was reviewed by George Odell and Bruce Huckell, both of whom made corrections and offered suggestions that greatly improved the end product.

Permission to use previously published illustrations was granted by editors Brad Lepper and Evelyn Lewis, respectively, of *Current Research in the Pleistocene* (Figs. 2.5 b–d, 2.6 a, 2.8 b, and 5.7 a), and *La Tierra* (Figs. 2.8 c and 2.9 a, b, c). The late Tom Kelly, author, and Richard McReynolds, artist, also graciously allowed us to use figures from their *La Tierra* article on the Van Autry Cores.

Marvin Kay acknowledges experimental control pieces supplied by George C. Frison, Anthony T. Boldurian, Susanne M. Hubinsky, Deborah Sabo, and George Sabo III. The automated photomicrographic system and the microscope used by Kay were purchased through grants awarded by the University of Arkansas Office of Academic Affairs and J. William Fulbright College of Arts and Sciences.

All of the professional staff at the University of Texas Press—especially Theresa May, Leslie Tingle, Lois Rankin, Ellen McKie—made production of this book a pleasant and rewarding experience. Tom Hester, too, as series editor, helped in many ways. Helen Simons did a wonderfully thorough job of copyediting. I thank all of these people for their efforts.

CLOVIS BLADE TECHNOLOGY

PART ONE
Blade Technology and Its Place in North American Prehistory

PRISMATIC BLADES OF STONE ARE CHARACTERISTIC of only a few prehistoric cultures in the Americas. The majority of those cultures are comparatively recent and also include pottery in their material culture inventory (Ford 1969). Sixteenth-century Spaniards were awed by the razor-sharp obsidian blades they encountered in their conquests of the Mayan and Aztec nations. Bernal Diaz del Castillo in his chronicles of Cortes's march to the City of Mexico in 1519 repeatedly described Indian weapons, including the "two-handled swords set with stone knives that cut better than our swords . . . so sharp . . . that they could shave their heads with them" (Idell 1956: 152), and remarked on the damage that these swords could do, as when "the Indians caught [Pedro de Moron's] lance so that he could not use it while others slashed at him with their swords and sliced at the mare [he was riding], cutting off her head" (Idell 1956: 100).

These Aztec and Mayan blades and their numerous prehistoric counterparts are part of a regional pattern of producing and working sophisticated polyhedral blade cores, mostly of obsidian but also of chert (Gaxiola and Clark 1989; Shafer and Hester 1983). Meso-american blade technology has been studied by a host of scholars who have amassed a great deal of archeological, ethnohistorical, and experimental data on where, when, and how blades were produced, as well as their economic importance and, in some cases, the social contexts of their production and consumption (Becker 1973; Boksenbaum et al. 1987; Charlton 1978; Clark 1982, 1984, 1985, 1987; Crabtree 1968; Gaxiola and Clark 1989; Hammond 1976; T. Hester 1978; Hester, Heizer, and Jack 1971; Hester and Hammond 1976; Hester and Shafer 1987; Hester et al. 1983; McSwain 1991; Nelson et al. 1977; Pires-Ferreira 1976; Shafer 1985, 1991; Shafer and Hester 1983, 1986, 1991; Sheets et al. 1990; Spence 1967, 1981; Stark et al. 1992; Tobey 1986).

This body of literature is the most voluminous and comprehensive treatment of blade technology in the Americas, but it is restricted to Mesoamerican blades that were generally produced by craft specialists at designated workshops in state-level systems of production and exchange (Coe 1994). Devices were employed by these knappers to gain sufficient mechanical advantage to detach blades by pressure (Clark 1982; Crabtree 1968). Other blade technologies in the New World were not as specialized, were part of less-complex cultures, and most tend to be less well known.

This book is concerned with Clovis blades, one of the lesser known of the blade technologies in the Americas and one that was practiced by hunters and gatherers near the beginning, rather than by state-level specialists near the end, of American prehistory. A need for a book like this became apparent when I began to study the Keven Davis cache of prismatic blades from Navarro County in northeastern Texas (Collins 1996). A few site-specific reports that touch upon Clovis blades have been published, but no synthesis has appeared.

Prehistorians concerned with Clovis manifestations in the early Paleoindian period of North America have not placed much emphasis on the blades and blade cores that were sometimes part of the techno-logical repertory of Clovis knappers. In recent years this aspect of the Clovis pattern has become somewhat more widely recognized and bet-ter documented. It would appear from recent developments that blades are particularly characteristic of Clovis in the southeastern and south-central areas of the United States and occur in minor numbers in the western parts of the Clovis range. Evidently, many Clovis localities in the northeastern part of the country lack blades entirely. Only very recently have archeologists working in the south-central United States given serious consideration to the presence and significance of blades in Clovis lithic assemblages. In reporting and interpreting blades from the Keven Davis cache, I attempt in this volume to add to the general understanding of Clovis blade technology while recogniz-ing that this effort is but a beginning and that it has several limitations.

One limitation of this study is its scope. I have focused on those data from the south-central United States with which I am most familiar. To these I have added selected information from other regions.

Another limitation, I suspect, is that there are many inadequacies in the reporting of Clovis blades in the literature. I am particularly interested in seeing how the presently perceived distribution of Clovis blades changes as awareness of blades increases. A few years ago, I doubt that half of the archeologists working with Clovis in Texas would have been aware of Clovis blades and blade cores in Texas sites, whereas now these artifacts are being recognized and reported with increasing frequency. Perhaps some of the areas that presently are blank on the Clovis-blade distribution maps will rapidly fill with reported finds as more archeologists become aware of the distinctiveness of Clovis blade technology.

Presumably with improved data on the occurrence of Clovis blades, understanding of the role blades played in Clovis adaptations also will improve. It is far from clear at this time why such blades seem characteristic of Clovis assemblages in many areas but are almost completely absent from Clovis sites in other areas. Equally puzzling is what appears to be the virtual absence of blade technologies in other Paleoindian manifestations.

Finally, there is the limitation that only a minor portion of what I consider to be the evidence on Clovis blade technology is derived from secure archeological contexts. Even the original definition of Clovis blades was based on specimens from disturbed context in the gravel quarry at Blackwater Draw (Green 1963). On the strength of Green's thorough documentation of the evidence and sound argument for those blades having been dislodged from the Clovis-age white sand, we have managed nicely for nearly thirty-five years with the improbable circumstance that this important aspect of Clovis technology was defined entirely on specimens from piles of loose dirt left by earth-moving machines.

Even today, we have only a few blades and blade cores from secure Clovis contexts, but these exhibit highly distinctive technological attributes. A greater number of specimens and assemblages are from less-secure contexts; these I infer to be also of Clovis affiliation because they manifest the same distinctive attributes as those of known Clovis affiliation. If we accept that inference for the moment, then the subjective interpretation of Clovis blade technology presented below can be proposed. Although it is a subjective interpretation, it can now be scrutinized because it has been articulated. Those archeologists interested in doing so should now go forth and subject my interpretations to full scrutiny.

Two prerequisites to the present consideration of Clovis blades are, first, to look closely at the nature of blades and the techniques of their manufacture and, second, to examine the full spectrum of Clovis lithic technology and how blade production articulates with the production of other stone tools. These prerequisites are addressed in Part One of this book. Part Two reports the Keven Davis blade cache in detail. Part Three presents a comparative study of blades, discusses caching behavior, and considers the implications that lithic technology and caching behavior have for the interpretation of Clovis lifeways.

TWO *Blades and Blade Technology*

Chipped-stone tools are produced by a reductive technology of controlled fracture. By applying an appropriate amount of force at the correct angle to a suitable spot on a piece of brittle stone, an artisan induces a fracture in the stone and detaches a piece of the desired shape and proportions. A sequence of successful detachments results in either shaping the parent piece to a desired form or accumulating several detached pieces with useful characteristics. This controlled fracturing is called *knapping* by students of stone chipping, and it is a human industry going back at least 2.4 million years (Schick and Toth 1993). Knapping has continued into the twentieth century in a few places in the world.

The most suitable kinds of stone for knapping are brittle, slightly elastic, silica-rich rocks or minerals that are *isotropic* (meaning that their optical, mechanical, and physical properties are the same in all directions) and have the property of *conchoidal* (smoothly curved or shell-like) fracture. Chert, flint, obsidian, quartzite, and basalt are common kinds of rock used by knappers.

At the most general level, the parent piece of stone is referred to as a *core* and the detached pieces are called *flakes*. When it is the core that becomes a finished implement, it is referred to as a core tool. Useful objects made on the detached pieces are called flake tools. Some stone-tool-making cultures emphasize the manufacture of core tools and dispose of most of the detached pieces as waste flakes. Others concentrate on the production of flake tools and treat the cores as waste. Most stone-tool industries, however, maximize the use of stone by making tools of both the cores and the suitable flakes. Whatever the strategy, any stone-knapping culture produces significant amounts of waste flaking debris, consisting of unusable flakes, small chips, irregular chunks, shatter, and ruined or exhausted cores.

Archeologists who specialize in the study of how stone tools are

made are commonly known as *lithic technologists*. In their study of the
multitude of diverse stone-tool cultures that have existed and the ves-
tigial ones that have continued into the modern era, lithic technologists have developed an extensive vocabulary of terms and repertory of
concepts. This has been done by close examination of tools and knapping debris from archeological sites, by documentation of the knapping behaviors of peoples in modern times who make and use stone
tools, and by the production of stone artifacts as controlled experimentation. Among general books on the topic, *Flintknapping* by
Whittaker (1994) and *Technology of Knapped Stone* by Inizan, Roche, and
Tixier (1992) provide good general introductions.

This book is concerned with the production of one particular form
of flake referred to as a blade. Blades and the technology of their production are specialized subjects in the study of lithic technology, and
they require introduction.

"BLADES" AND "TRUE BLADES"

The distinction between blades and other flakes has long suffered
from lack of total agreement among archeologists, and several definitions have been proposed. A minimal but widely used definition is that
put forth by the eminent Old World prehistorian and lithic technolo-
gist François Bordes, classifying as a blade any detached piece more
than twice as long as it is wide (Bordes 1961). That definition includes
any kind of flake that happens to be of those proportions (Fig. 2.1 a).
The context of Bordes's definition is often overlooked by New World
specialists. He was defining the term for use in classifying Lower and
Middle Paleolithic stone tools, where blades by any definition are relatively infrequent (except in a few localized instances). In contrast,
during the Upper Paleolithic, blades—often called "true blades"—are
far more common and they meet more stringent definitions, even in
Bordes's own writings (Bordes 1967; Bordes and Crabtree 1969) where
emphasis is placed on the technique of production, not just the proportions of the piece. Bordes and Crabtree conclude from archeological and experimental evidence that Upper Paleolithic blades ("true
blades," Fig. 2.1 b) were produced in three basic stages and retain the
attributes imparted by this mode of production. Their three stages are
succinctly summarized by Newcomer (1975: 99) as follows:

> *the preparation of a striking platform at one or both ends of the core,*
> *the "cresting" of the front of the core by bifacial flaking to provide a guiding*
> *ridge for the removal of the first blade, and the detachment of blades, with*
> *platform preparation between blade removals to remove overhangs left by the*
> *negative bulbs of previously struck blades.*

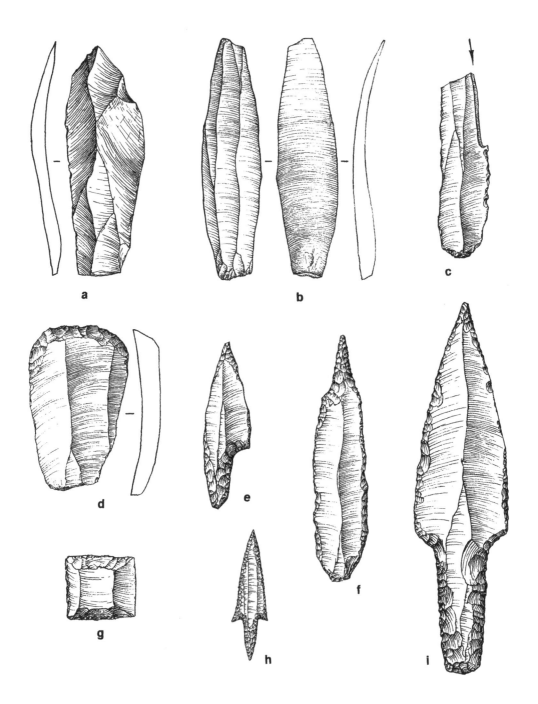

Fig. 2.1 Blades and various tools made on blades or blade segments: (a) "blade" as defined by Bordes (1961); (b) true blade; (c) Upper Paleolithic burin; (d) end scraper; (e) Solutrean *pointe a cran;* (f) perforator; (g) gun flint; (h) Perdiz arrowpoint; (i) Mayan "dagger." (Not to scale.)

Blades produced in this manner have several distinctive attributes. They are almost always significantly more than twice as long as they are wide; those from the Upper Paleolithic site of Corbiac in southwestern France, for example, generally exceed length-to-width ratios of 4 to 1. Except for initial, *crested*, blades (see below), these blades have subparallel blade facets on their exterior surfaces (Fig. 2.1 b). They are triangular, prismatic, trapezoidal, or trapezium-like in cross section and have relatively even lateral margins. Generally, platforms are small, bulbs are not prominent, and interior surfaces are nearly smooth (Fig. 2.1 b). In Crabtree's (1982: 16) words, a blade is a

> *Specialized flake with parallel or sub-parallel lateral edges; the length is equal to, or more than, twice the width. Cross sections are plano convex, triangulate, subtriangulate, rectangular, trapezoidal. Some have more than two crests or ridges. Associated with prepared cone [sic; read core] and blade technique; not a random flake.*

Crabtree (1982: 36) further defines flake as follows [uncorrected]:

> *Any piece of stone removed from a larger mass by the application of force, either intentional, accidentally or by nature. A portion of isotropic material having a platform and bulb of force at the proximal end. The flake may be of any size or dimension, depending on which technique was used for detachment.*

Specialists, particularly those working with Upper Paleolithic assemblages where blades are one of the defining traits, have clearly based the definition of blades, meaning true blades, on the mode of production, thereby eliminating elongate flakes produced by chance. For the remainder of this study, the term *blade,* unless otherwise qualified, is used in reference to those forms intentionally and systematically produced by a "blade technology."

BLADE TECHNOLOGY

Blade technology refers to the knowledge, strategy, material, activities, and equipment involved in the intentional production of true blades. Skill and fairly involved procedures are required, but the effort is warranted because of the beneficial characteristics of blades. Structurally they are quite strong, and they have regular, smooth, and very sharp edges. This combination of attributes allows versatile applications, and the archeological record is rich with artifact forms made on blades or segments of blades (Fig. 2.1). Segmentation of blades into rectangular or other geometric shapes is a common technique for obtaining uniform pieces for use or for further modification. Blades and blade segments can be used unmodified as cutting tools or they can be retouched into various cutting, piercing, or scraping tools (Fig. 2.1 c, d, f, i); more specialized forms include projectile points and gunflints (Fig. 2.1 e, g, h).

Blade technology is the most efficient use of stone in terms of total length of cutting edge produced from a given mass of stone (LeRoi-Gourhan 1943; Sheets and Muto 1972; Whittaker 1994). A technology that emphasizes the production of flake tools is somewhat less efficient, and a core-tool technology is the least efficient.

A common variant of blade technology emphasizes the production of small blades, termed *microblades* (cf. Tixier 1963). Prehistorians generally define microblades as being less than 3 cm in length (de Heinzelin de Braucourt 1962: 14; Brezillon 1968: 100) or as being less than ca. 1 cm in width (Owen 1988: 2). In this study of Clovis blades, which are full sized, we are not concerned with microblade technology, although some of what is said here applies equally to microblades.

An unfortunate ambiguity exists in the literature, where another term, *macroblade,* is occasionally used to refer to full-sized blades (as distinct from microblades) of the sort common to the Upper Paleolithic and to Clovis (e.g., Baumler and Downum 1989: 112) but by others to identify extraordinarily large blades, especially those greater than ca. 15 cm long found in some knapping sites in Central America (T. Hester 1972; Shafer and Hester 1983; Soto de Arechavaleta 1990). In a majority of works on both Mesoamerican and Paleolithic assemblages, the unmodified term *blade* refers to normal-sized blades (generally between 3 and 15 cm long) which is the usage followed here.

Archeologists in the Americas describing the artifacts from a site will often enumerate "blades" (e.g. Johnson 1991) using Bordes's (1961) definition. As a loosely descriptive category, this may be justified, but it does not necessarily follow that these are the products of a true blade technology as I have defined it above. Intentional and systematic production of true blades will leave a distinctive archeological record composed of blades, blade cores, exhausted blade cores, and an array of characteristic debris from the reduction and maintenance of blade cores (Fig. 2.2). Often these products and byproducts of blade technology are found in staggering quantities, as for example the large mounds of obsidian debris from blade manufacture at the site of Sierra de Navajas in Hidalgo, Mexico (Nieto Calleja and Lopes Aguillar 1990:196). When these copious data are examined carefully, the details of each distinctive blade technology emerge and can be precisely defined.

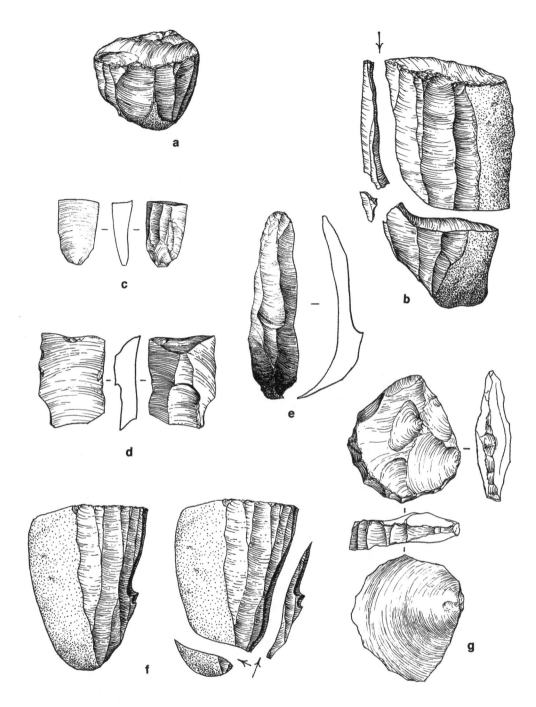

Figure 2.2. Typical kinds of knapping debris resulting from blade production: (a) depleted core; (b) core ruined by diving blade; (c, d) blade fragments; (e) hinge-recovery blade with excessive curvature; (f) core successfully redeemed by removal of a hinge-recovery blade from a secondary platform prepared for that purpose at the distal end of the core; and (g) a core tablet flake.

Various specific blade technologies have been so defined around the world, including the Upper Paleolithic of Europe, the Near East, parts of east and south Africa, northern Asia, and Japan (Bordes 1968). These share the application of basic blade-production principles (see below) but they are distinguished by four key variables, some cultural and some imposed by nature. These variables are:

the size, form, quality, quantity, and lithology of available raw materials,
the intended size and form of the blades produced,
the reduction strategy and techniques employed, and
the tools that the knapper uses.

There are any number of permutations to each of these variables, and the combinations make for an almost infinite array of variations in blade technologies. In spite of this, there are some basic requirements common to all blade production technologies that are governed by fracture mechanics.

Raw material must be relatively homogeneous, isotropic, and available in large enough pieces for blades to be produced. In any knapping, because it is reductive, the size of the objects produced is necessarily less than the size of the parent piece of raw material. Since blade production is often concerned with maximizing length of the blades, length is the only dimension limited by raw material sizes. Blades can almost never be as long as the greatest dimension of the raw material, because some portion of the material, especially a suitable platform at the core's upper end, must be sacrificed in forming the core. It is on the platform that force is directed in the detachment of blades. Homogeneous, isotropic, brittle stone responds to force with even, predictable fracture, allowing blades of the desired proportions to be produced. Stone lacking in any one of these characteristics may serve well for some knapping but is seldom suitable for the production of blades. Homogeneous, isotropic quartzite, for example, may not be brittle enough for blade production, because of the great force needed to bring about fracture; brittle, isotropic obsidian with internal flaws or fractures may not flake predictably enough for blade production; or brittle, homogeneous quartz with anisotropic crystalline structure is almost never suitable for blade production. Because of limitations in properties of stone, the major blade-producing technologies of the world are all centered on abundant sources of superior chert or obsidian. Even in those favored localities, the vagaries of natural stone materials are evident in numerous failed attempts at blade removal.

Chippable stone fracture is controlled by the form of the parent piece, the angle and amount of force applied, and the nature of the material used in applying the force. Optimally, conchoidal fracture results in the removal of a flake with thin, sharp edges and leaves a smooth, slightly concave depression in the face of the core (the flake scar). This happens when force is directed beneath a slightly convex surface on the core face; if that surface is flat or at all concave, the flake will usually terminate prematurely with a thick, dull edge and the face of the core will possess a stepped flake scar. These are almost always undesirable results. If the face of the core to be flaked is simply convex, as in the exterior of a rounded stream cobble, the flake detached will tend to be nearly round with its width almost as great as its length. If, however, there is a ridge down the face of the core, the force will follow beneath that ridge and detach an elongate flake— usually oblong, unless special conditions exist to make it even longer. It is just those special conditions that must be in place to produce blades. The face of the core must be at least slightly curved outward and there must be a ridge (or crest, see below) to direct the force down the face of the core. To achieve blade proportions, the knapper's force must be precisely directed into the material beneath that ridge and the platform receiving that force must meet certain specifications.

Blade knappers are concerned with the angle and amount of force they deliver into the mass of a core, as well as with the precise point at which that force is directed. Like the golfer's swing or the diamond cutter's blow, the success of the blade knapper's fleeting application of force is largely determined by a series of preparatory adjustments and planning. Because failure and frustration abound in this process, I suspect that in a remote time when stone tools and language were nascent human traits, knapping was the mother of profanity.

Blades may be detached by striking the platform of the core with a sharp blow; this is called *direct percussion*. Direct percussion may be with an object as hard as, or harder than, the core material. In this case the object is referred to as a hammer and the process is called hard-hammer percussion. Or, direct percussion may be with a material softer than that of the core—soft-hammer percussion. Soft hammers are often of antler, wood, or bone and are commonly referred to as billets or batons. Some knappers use soft stone hammers for this purpose. An important aspect of direct percussion is that the knapper must stabilize the core, usually with one hand, and deliver the blow at precisely the right spot on the platform, at the correct angle, and at the correct velocity. This is usually done with the hammer or billet in the other hand (Fig. 2.3). All of this requires good hand and eye coordination. In a comparatively rare version of direct percussion, the core is swung against a fixed object, such as a promontory on a large rock, an operation equally demanding of the knapper's skill and coordination.

Percussion force may also be delivered through an intermediate

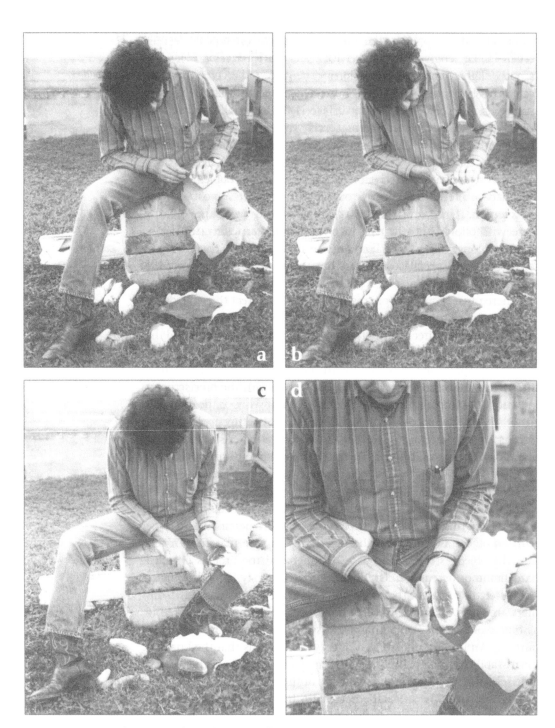

Figure 2.3. Glenn T. Goode producing blades by direct percussion: (a) platform preparation by direct percussion with a small hammerstone; (b) abrading the platform with the same hammerstone; (c) direct-percussion blow just completed; and (d) blade successfully detached. Note the array of knapping tools around Goode's feet and the position he uses in holding (steadying) the core behind his leg.

object, called a punch or drift, that is held with one end in place on the platform and then struck, usually with a mallet or soft hammer, at the opposite end. This is called *indirect percussion*. It has the advantage over direct percussion that the punch can be placed and held very precisely at the correct angle and exactly on the desired spot on the core. The disadvantage in most cases is that the punch is held in one hand and the mallet swung with the other, which requires some means of holding the core securely with the feet, legs, or in a holding device of some kind (Fig. 2.4). Sometimes a second person is needed to hold the core for the knapper. The demands of skill and coordination are at least as great as in direct percussion.

Finally, force can be delivered in the form of pressure. In *pressure flaking*, a pointed tool is carefully placed on the platform and a force load is increased by the knapper until fracture is initiated in the core. The force may be generated by the knapper's own strength and body weight on a pressure tool designed for that purpose, sometimes called a crutch. As in indirect percussion, some means of stabilizing the core with the feet, in a device, or with the assistance of another person is necessary, since both hands are used in guiding the pressure tool and generating the force. Leverage may be substituted for, or combined with, body strength and weight. Pressure is suitable only for the production of microblades or normal-sized blades in extremely brittle, homogeneous, and fully isotropic material—almost exclusively obsidian. Pressure blades of obsidian are the pinnacle of the blade maker's art in size, length-to-width proportions, keenness and sharpness of edge, and uniformity. This is because the position, angle, and amount of force can be very precisely controlled. But the specifications and maintenance of the pressure tool, the preparation and maintenance of the core, the need for an abundance of high-quality stone, and the technology for securely holding the core all require great effort, as well as considerable skill and training. Not surprisingly, most archeological evidence points to pressure blades as being the work of full-time craft specialists.

For blade making in general, the combination of variables at work is almost infinite, and it would serve little purpose to attempt further discussion that takes all permutations into account. Instead, since this discussion of blade technology is a prologue to the study of Clovis blades, it is appropriate to focus on the production of blades of that general kind, which would also apply to most of those of the Upper Paleolithic. This can be done by presenting the basic features of an idealized blade technology similar to that used in the Upper Paleolithic of Europe and in Clovis culture.

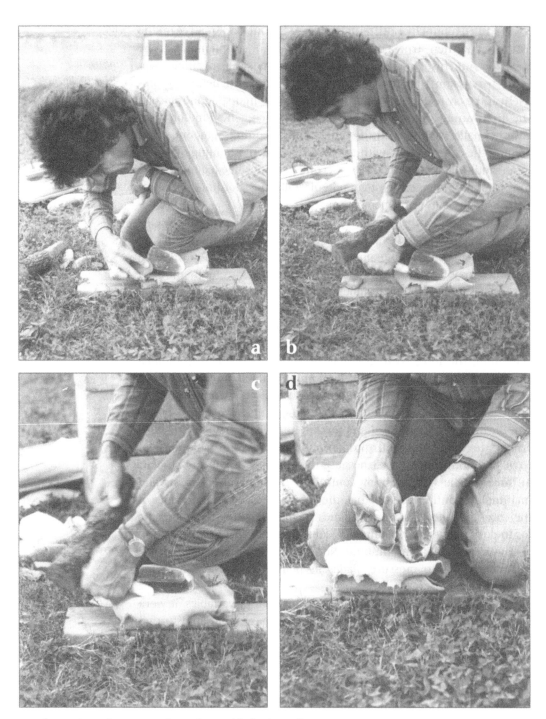

Figure 2.4. Glenn T. Goode producing blades by indirect percussion: (a) grinding the edge of the platform with a hammerstone; (b) punch in place on the platform of the core while a practice swing is made with the mallet; (c) blade in flight at the instant of detachment; (d) blade successfully detached. Note the mass of the wooden mallet and the dull tip of the antler punch; note also the technique of holding the core under the knee.

To present and illustrate the essentials of blade technology, this discussion describes a sequence of steps that a knapper might take in producing blades similar to those found in Clovis or in Upper Paleolithic sites. Where appropriate, significant optional or alternative procedures are considered. For each step, the nature of the material being worked is described, the objective of the step is explained, the means of obtaining the objective are discussed, and the resulting products—both intended and incidental—are described. This presentation is simplified by arbitrarily limiting it to a discussion of chert knapped by direct or indirect percussion.

This discussion draws on my own knapping experience, observations of, and discussions with, expert knappers (F. Bordes, B. Bradley, J. E. Clark, D. Crabtree, J. Flenniken, G. Goode, B. Huckell, H. Nami, D. Stanford, G. Titmus, and J. Tixier), and numerous published sources (specifically Bordaz 1970; Bordes 1947; Bordes and Crabtree 1969; Callahan 1979; Clark 1982, 1985; Clarke 1935; Cotterell and Kamminga 1987; Crabtree 1968; Crabtree and Swanson 1968; Ellis 1939; Flenniken 1985; Inizan, Roche, and Tixier 1992; Luedtke 1992; Newcomer 1971, 1975; Owen 1988; Sheets and Muto 1972; Speth 1972; Whittaker 1994; Wilke and Quintero 1994).

The first task in any knapping process is to acquire suitable raw material. Suitability in this case means that the chert is in large enough pieces, is of sufficiently good quality, and either occurs naturally in, or can be readily transformed into, shapes that allow normal-sized blade removals.

Chert occurs in limestone either as irregular, rounded nodules or as beds. The knapper must find where such stone occurs and select useful pieces. It can be quarried from the bedrock, found as loose pieces weathered from bedrock, or gathered as stream cobbles transported some distance from their point of geologic origin. Bedded or ledge chert is usually blocky (Fig. 2.5 a) whereas nodules or cobbles are commonly rounded (Fig. 2.5 b). Any piece of chert—block, nodule, or cobble—may have a natural shape conducive to blade removal with little preparation; that is, it may have one or more ridges and a natural platform (Fig. 2.5 a, c). More often, however, some shaping will be required to form a core. It is prudent to test chert at the place where it occurs to be sure that is does not contain unseen flaws. The entire knapping process may transpire at the source of the chert, or unmodified raw material may need to be transported to the knapping area; typically, however, some preliminary reduction is done at the source and the job is completed at another locality.

Chert is found with a weathered exterior surface called the cortex, which in most cases, is soft and desilicified. Cortex is not useful in

Figure 2.5. Typical occurrences of chert: (a) blocky piece of ledge chert; (b) ovoid nodule; (c) nodule with a natural flat surface suitable as a blade-core platform with little or no preparation. (Not to scale.)

most cases and must be removed in the knapping process to expose the hard interior chert. If the cortex is thin, it does not interfere with knapping, but thick, spongy cortex can absorb as well as misdirect the force of a blow and may need to be cut away with a flake or other tool before knapping can begin.

Following the acquisition of raw material, the knapper's next step is to begin reduction of the naturally occurring core (Fig. 2.6) or to prepare a blade core for reduction (Fig. 2.7). The objective in this step is to establish the required attributes in the core for successful removal of a series of blades. Often the first several blades are not usable, but the scars they leave on the face of the core are critical to the success of further removals, so the knapper must consider the full potentiality of each piece of raw material when preparing for the first blow. Blocky ledge cherts often possess an angular edge that can be used to direct the first blade. If this edge is oriented properly in relation to a corner, no preparation of a platform may be necessary. Nodules and cobbles are often shaped somewhat like a large potato (Fig. 2.6 a), and the knapper can shape them into suitable blade cores simply by knocking off one end (Fig. 2.6 b). The resultant fracture surface serves as the platform, and there is usually at least one elongate, convex face where the first blade can be removed (Fig. 2.6 c). The two ridges on the face of the core along the edges of that first blade scar become the guiding crests for subsequent blade removals.

If a core lacks any suitable natural face for the removal of the first blade, a ridge can be produced by bifacial flaking (Fig. 2.7). This is referred to as a *crest,* and the detached blade, which is triangular in cross section with the prepared crest as its exterior, is termed a *crested blade.* Some knappers will establish platforms at both ends of a blade core at this stage, but generally this occurs only in more specialized blade technologies. In addition to a core and perhaps some useful early-stage blades, some number of waste flakes, many with cortex on their exteriors, will be produced at this stage. The first blade from an unprepared core will be entirely covered in cortex (Fig. 2.6 e), several successive blades will have partial cortex (often along one edge [Fig. 2.6 f, g]), and subsequent ones may have small remnants until the cortex is all removed from the core. Some raw material has a partial natural ridge that requires only partial cresting, so crested blades may have a flaked crest for all or only part of their length. Ultimately the cores (Figs. 2.6 d and 2.7 f) and the blades (Figs. 2.6 g, h and 2.7 i, j) from both approaches will be indistinguishable.

CORE PREPARATION AND INITIAL REDUCTION

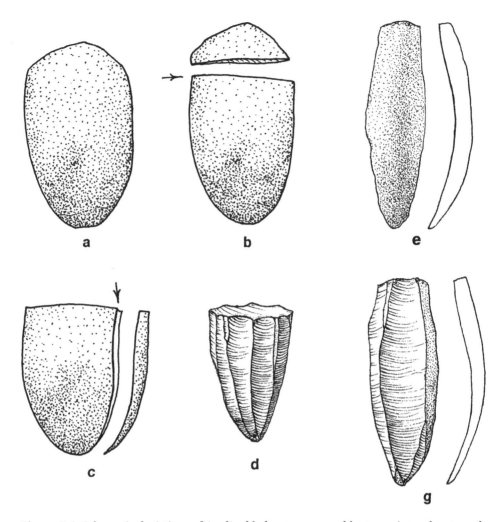

Figure 2.6. Schematic depictions of (a–d) a blade core prepared by removing only one end of a nodule and using the natural contour of the nodule as the core face, and (e, g) typical blades resulting from this strategy for core reduction.

PLATFORM, THE ANGLE OF FLAKING, AND CORE MAINTENANCE

At this point the particulars of platform preparation and angles of flaking must be explained, before discussion of the next knapping step. As alluded to repeatedly above, these are critical requirements in any knapping endeavor and must be considered by the knapper throughout the reduction of a blade core. A core is shown in schematic section in Fig. 2.8 a with its platform at the top and its face to the right. When a blade is detached, a portion of the core face and of the core platform go with the blade; the detached sector of the core

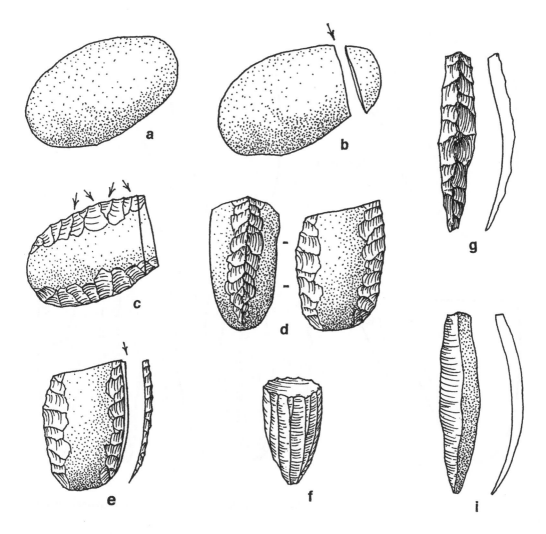

Figure 2.7. Schematic depictions of (a–f) a nodule prepared by removing one end and by flaking a crest on two faces to guide the initial blades, and (g, i) typical blades resulting from this strategy for core reduction.

face becomes the exterior surface of the blade and the detached portion of the platform is still referred to as a platform (Fig. 2.8 d). Both of these surfaces of the blade retain attributes imparted to the core before detachment of the blade. The interior of the blade and the blade scar it left on the core face (Fig. 2.8 d) have characteristics imparted during detachment, of which more will be said later.

Force is directed onto the platform of the core (whether through direct or indirect percussion) at a given angle to the platform and at a

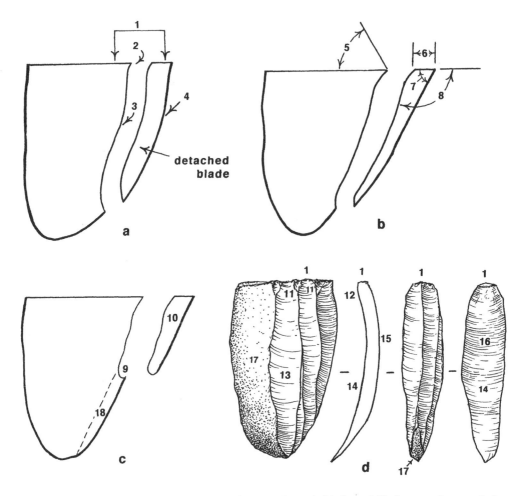

Figure 2.8. Basic blade and blade-core characteristics.

specific point back from the edge of the platform. This angle is referred to as the *angle of flaking* (Fig. 2.8 b) and the distance from the edge of the core defines the *platform depth* on the detached blade (Fig. 2.8 b). An all-important relationship exists in the angle between the face and the platform of the core, or the *exterior platform angle* (Fig. 2.8 b). This angle must be acute, but the closer it is to 90 degrees, the longer the detached blade will be, other variables being constant. Optimal values for the angle of flaking and the platform depth vary with material, mass of the core, mass of the percussor, exterior platform angle, condition of the core face, and other subtle factors. A knapper depends upon experience and judgment in simultaneously weighing all of these variables, some consciously and others subconsciously. If the angle of flaking is too shallow, a short flake will be knocked off and may ruin the face of the core; if the angle is too steep, force will dissipate into the mass of the core, also potentially ruining the piece.

As each blade is removed, its interior surface near the proximal end will manifest a rounded, slight eminence known as the *bulb of percussion* (Fig. 2.8 d); a corresponding depression in the face of the core just below the platform is variously referred to as the negative bulb of percussion, negative bulb scar, or negative bulb, or *bulbar scar* (Fig. 2.8 d). This negative bulbar scar creates an overhang at the juncture of the core face and platform (Fig. 2.8 a). It is essential that this overhang be removed before the detachment of another blade is attempted. This is ordinarily done by taking off small flakes to chip away the overhang, part of a process called *core maintenance*. If the concavity of the negative bulbar scar is not eliminated through removal of this overhang, the next attempt at blade removal will only detach in a short flake with a hinged or stepped termination that ruins the core face (Fig. 2.8 c). It is sometimes possible to restore the core face to a usable condition by detaching a *recovery flake*, which is usually struck from the base of the core, in the opposite direction as the failed blade removal, or from one side of the core, perpendicularly or diagonally to the failed blade removal. This is another aspect of core maintenance.

Two additional core-maintenance tasks are routine in blade making, one to keep the blades from becoming too curved and the other to restore the required attributes to the core platform. As a series of blades is removed from the same face of a core, each blade may slice away more of the distal end of the core, and subsequent blades become increasingly curved in longitudinal section. Too much curvature is often undesirable in blades and if not held in check, increasing curvature of blades can eventually lead to a blade plunging into the mass of the core and detaching a large part of the base of the core (Fig. 2.2 b). One common procedure for correcting this excessive curvature is to strike blades intermittently from the opposite end of the core. Another is to increase the angle of flaking.

As blades are removed from a core (Fig. 2.9), the face of the core is constantly changing, as is the relationship between the core face and the platform; damage to the platform sometimes occurs. Minor flaking of the core face or platform will compensate for some of these changes, but eventually the configuration of the core reaches a point beyond which no more blades can be removed. However, if sufficient mass is left in the core to warrant it, the core can be made useful again by completely removing the platform as a single large flake, known as a *core tablet flake* (Fig. 2.9 d). The peculiar practice of platform maintenance used by Clovis knappers resulted in the need for frequent core tablet removals (see Chapter 3).

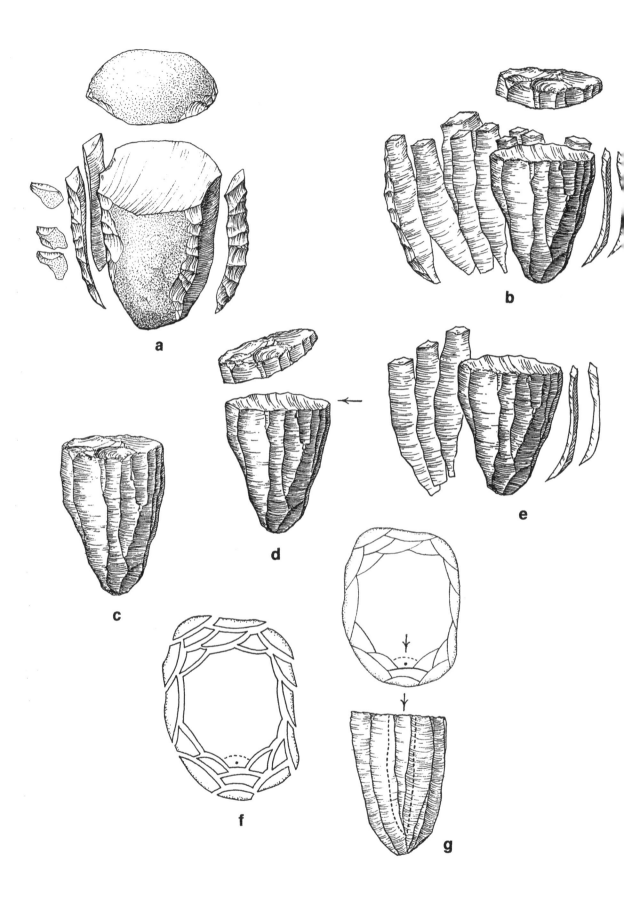

a

b

c

d

e

f

g

Figure 2.9. Schematic depiction of a typical blade-core reduction: (a) early stage of roughing out the core and removing crested blades; (b) core, blades, and core tablet flake; (c) core with dysfunctional platform; (d) core tablet removed; (e) blades detached subsequent to successful removal of core tablet; (f) idealized depiction of the pattern of detached blades as seen from above the platform, showing why on the core face only the scars of the two shaded blades are as wide as the blades and how the next blade removal (indicated by the dashed line) will alter that pattern; (g) the detachment of another blade as shown is anticipated to alter the core face as shown.

CORE REDUCTION AND BLADE PRODUCTION

Having configured the core and removed some preliminary blades to set up ridges to guide subsequent blades, the knapper is ready to reduce the core and produce a series of blades (Fig. 2.9). If the angle between the platform and the face of the core is near 90 degrees, it may be possible for the knapper to move around the entire circumference of the core, removing blades in a spiral pattern until the core is depleted. Of course, this can be done only if the platform and core face are rigorously maintained and no blade detachment failures occur that cannot be recovered. If the core is irregular or the platform is at an acute angle to the core face, the sequence of removals will sweep back and forth across a core face that is limited to one side of the piece of raw material.

For each blade removal, the artisan identifies the precise point on the platform, determines the angle of flaking, and steadies the core and holds it firmly. Identifying the point on the core platform for directing force entails a consideration of the core face (aligning with a ridge) and details of the platform (its strength and angle largely dictate the depth of the platform to be detached). In the case of indirect percussion, a mallet (some knappers use their billet) is used to strike the punch that is held at the correct angle of flaking and that conducts the force into the mass of the core. One of the most common problems encountered at this point is slippage of the punch on the platform surface. This risk can sometimes be reduced by roughening the platform surface with an abrader.

In the case of direct percussion, the billet is swung against the platform in an arching path that makes contact with the platform in a glancing blow; experienced knappers know the interplay between the speed of the swing, mass of the billet, mass of the core, configuration of the core, and angle of that glancing blow to produce force equivalent to a straight-line force at the optimal angle of flaking. The action is somewhat analogous to a pool shot where the cue ball traveling in one direction strikes another ball with just the right amount of "bite" and force to propel it at the desired angle to the path of the cue ball.

A hazard common to both direct- and indirect-percussion core reduction is shifting of the core under the knapping force. When this occurs, it almost always results in a failed blade removal and may cause severe damage to the core. A very slight amount of movement of the core may occasionally be desirable to absorb some of the shock of the knapper's blow, but this is distinct from unintended slippage or rotation.

If no failure occurs, a blade is successfully detached and the process can be repeated until the core is depleted.

The more precisely force can be directed in terms of the angle of flaking and the point of impact, the more successful blade detachment is likely to be. It is often beneficial to remove small guiding flakes on both sides of the ridge on the core face to isolate and make more prominent the targeted spot on the platform.

Blades will become progressively more uniform if core reduction proceeds as intended. Edges will be straighter and cross sections will be more uniformly prismatic. Little or no cortex will remain on more-interior blades. Of course, with each platform rejuvenation, the core and any subsequently detached blades become shorter. This will be especially true if platforms are being maintained at both ends of the core. A distinctive part of the sequential removal of blades is that each new blade overlaps the scars of previous blades. This gives the prismatic form to the blade cross section. It also means that the scars of all but the last blade in a sequence are partially removed, leaving scars that are narrower than the corresponding blades, a fact sometimes overlooked by analysts tempted to use blade scars on cores to estimate blade widths.

Each blade technology will operate with standard range of lengths and widths for the blades it produces. When a core is depleted to the point that suitable blades can no longer be obtained, the core is discarded. Cores may be discarded at earlier stages of reduction if a flaw in the material is encountered or an unrecoverable knapping failure occurs.

Usable blades along with blade fragments, ruined or depleted cores, core fragments, core tablets and other core maintenance flakes, and small flaking debris are all produced in the course of reducing a blade core. The usable blades may require modification before being put to use.

BLADE MODIFICATION

Detached blades may be used as detached or they may require modification. Typical modifications are segmentation and truncations. *Segmentation* into shorter pieces is easily accomplished by striking the blade on the interior or exterior face and causing it to break. This produces abrupt fracture planes at the ends of the segments. An alternative to segmentation is to chip the blade into segments, a process referred to as *truncation,* which produces a flaked end to the blade segment. Truncations are often in the form of scraper edges, but fracture scars or truncations may be used as platforms for detaching small flakes along the blade edge to produce chisel-like tools called burins. Any number of other tools can be formed by flaking blades or blade segments into desired forms (Fig. 2.1).

EXPERIMENTAL REPLICATION

One of the most powerful analytical tools used by lithic technologists in the study of prehistoric stone tool assemblages is experimental replication. A number of lithic technologists have replicated prehistoric blade technologies (e.g. Bordes 1947; Bordes and Crabtree 1969; Clark 1982, Newcomer 1975; Sollberger and Patterson 1976; Whittaker 1994; Wilke and Quintero 1994) and their findings contribute significantly to our understanding of past behaviors. One must always be cautious of over-interpreting replication results, because it is possible to arrive at similar products using different techniques. So, successful replication means only that the same techniques *could* have been used in the past. Independent evidence is needed to assess how closely the replicator's knapping tools, techniques, and strategies might resemble those used in prehistory. But replication always affords insights into the details of any technology that are not likely to be discovered by observing only prehistoric evidence.

This study of Clovis blades has benefited greatly from the replicative work of Glenn T. Goode of Austin, Texas, who can produce blades very similar to those found in Clovis contexts. Goode generally uses direct percussion (Fig. 2.3) but also makes blades using the punch technique (Fig. 2.4). His blades (Fig. 2.10 b, i) share the distinctive Clovis attributes of strong curvature, small platforms, smooth blade interiors, and small bulbs of percussion. His cores (Fig. 2.10 d, f, h), however, are usually of the wedge-shaped form that are less common to Clovis than are the conical forms. He seems to prefer a platform angle of about 75 degrees. When he occasionally makes and reduces a conical core, it closely resembles those of Clovis origin (Fig. 2.10 e, f).

Goode has produced a few thousand blades from a few hundred blade cores over the past six years (spring 1991 to spring 1997). His store of knowledge and stockpile of successful and failed cores and blades are invaluable resources to anyone interested in blade-making technology.

Goode works primarily with cherts from the Edwards limestone formation, especially the widely revered "Georgetown" variety, when making blades. His knapping tools typically include four or five hammerstones of various sizes and materials, three or four antler billets, a wooden mallet, several antler punches, a few antler pressure flakers,

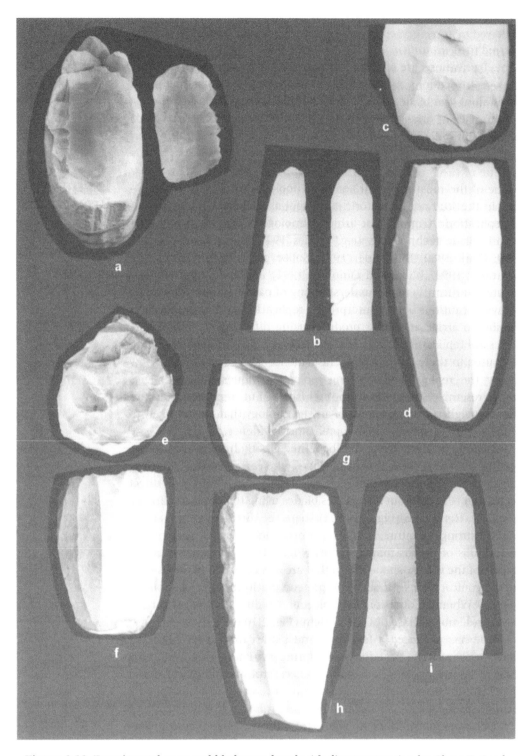

Figure 2.10. Experimental cores and blades produced with direct percussion by Glenn T. Goode: (a) oblique view of a rejuvenated core platform, the interior of the freshly detached core tablet, and three refitted blades showing how the negative bulb scars of those blades have been removed from the core face; (b–d and g–i) blades and blade core illustrating variation in platforms on blades, typical core-platform preparation, and negative bulb scars present on core face since the core platform has not been rejuvenated; (e, f) blade-core face and platform closely resembling specimens from prehistoric contexts.

and some leather padding (Figs. 2.3 and 2.4). He uses no holding devices to secure the core, holding it in one hand against the inside of his leg for direct percussion or under his knee for indirect percussion (Figs. 2.3 and 2.4). He has experimented with other holding positions, such as placing the core under his foot or having the core held by an assistant (usually Robert J. Mallouf), but for most purposes finds the positions illustrated here the most satisfactory. He does most of his platform maintenance with the hammerstones, employing both chipping (Fig. 2.3 a) and abrading techniques (Figs. 2.3 b and 2.4 a). On occasion, it is necessary to detach small core-maintenance flakes with a pressure flaker or to use the antler billet to take larger flakes off of the core face or platform.

Goode commonly follows the strategy of severing one end from a rounded nodule to form a platform and proceeding to remove (cortical) blades with no preparation of the core face. Alternatively, he selects nodules that have a suitable platform that formed in nature along a plane of weakness in the stone (e.g., Fig. 2.5 c). After removal of the primary cortex blade and several secondary cortex blades, the core begins to yield regular blades lacking cortex. His maintenance of platforms (e.g., Fig. 2.10 c, e, g) commonly results in highly faceted platform surfaces that may eventually have to be removed as a tablet (Fig. 2.10 a).

Since beginning in the spring of 1991 to replicate Clovis blades, Goode has made many very close replicas in terms of all of the Clovis blade attributes, including size. However, most of his effort has been devoted to reduction of cores that he can hold without any support other than his leg. To replicate the largest Clovis blades requires cores just a little too big for Goode to hold in this fashion, so most of his experiments have produced pieces in the lower 75 percent of Clovis blades in terms of length. When he does produce blades in the upper range of lengths, supporting the larger core on a block of wood or on the ground is often required.

Goode is able to get twenty or more blades from a single core if all goes well, but he rarely does, for several reasons. Trying to get a large number of blades from a single core has never been a goal for Goode, and the question of how many might be obtained per core has not yet been addressed in his experimentation. Since his usual objective in replication is to better understand past behavior, he often encounters a specific condition in the reduction of a core that he wants to preserve for further examination; in such cases, he sets the still-viable core aside with no more reduction. He estimates that he may average ten or slightly fewer blades per core, given the constraints imposed by other research objectives.

Since most blade makers in prehistory went to considerable effort to make their blades as straight as possible, the strong curvature often seen in Clovis blades raises some interesting questions. When I ask Goode about making less curved blades, he responds that he does not put much effort into making straight blades, since he is replicating the curved form seen in Clovis blades. Even so, he explains that increasing the angle of flaking to near 90 degrees and reducing the platform depth contribute to making straighter blades, as does frequently reversing the core to take blades off in opposing directions from the same face of the core. He also notes that blades tend to be straighter when the core is seated on a firm support at the time of blade removal, all else being equal.

In Goode's experience, supporting the core on a firm base—a block of wood or the ground—has resulted in higher incidence of blade breakage and blades with less refined attributes. Although blades are straighter, their edges are less regular, bulbs are larger, and interior surfaces have more pronounced ripple marks.

This brings the discussion to one of the more distinctive and intriguing aspects of Clovis blades—their smooth interior surfaces with small bulbs of percussion and minimal ripple marks. In his efforts to replicate this trait, Goode has found that, up to a point, the less firmly a core is supported, the smoother the blade interiors will be. Of course, if supported too loosely, the core will recoil under the knapper's blow and dissipate the force before fracture is complete, likely ruining the core. In holding the core, two requirements must be met. First, the core has to be absolutely steady so that the knapper can precisely deliver the blow. Second, the core has to yield under the blow ever so slightly (see below) but not enough to interfere with detachment. This minute bit of yield is so slight as to be beyond the knapper's control.

As noted, Goode has detached blades with the core resting with its base on the ground or on a block of wood, and both of these techniques resulted in blades with larger bulbs and more rippled interior surfaces, as well as in more blades breaking during removal. For a time, he supported the core with its base against a thick pad on the inside of his leg, driving the blades off directly toward the pad. The results were better than when the core was on the ground or a block, but still included many broken blades and blade interiors less regular than those seen on Clovis specimens. In these arrangements, the support was directly in line with the direction of force. By moving the core back to a position slightly behind his leg with the base not supported against anything (Fig. 2.3 c), Goode could hold the core steady and he could hold it firmly enough in his left hand for blade detachment. With the force directed toward the open space behind his leg, he can routinely obtain blades with smooth interiors and fewer breaks.

This discussion has repeatedly mentioned bulb size. In this context, bulb size is relative. We are considering the variation in sizes among soft-hammer and punch techniques, not between hard- and soft-hammer techniques. Virtually all hard-hammer blades will have larger bulbs of percussion than even the largest found on soft-hammer blades.

The very small platforms seen on Clovis blades are replicated by Goode in direct percussion simply by very accurate, low-angle blows with a soft hammer to a strong, well-isolated platform. The face of the billet may be several centimeters across, but it is rounded, and a very small area actually contacts the core. By the same token in indirect percussion, Goode uses fairly dull punches with rounded ends so that only a small area on the rounded tip of the punch is in contact with the core. More-pointed punches can ensure small platforms, but they are more subject to splitting and require much more frequent refurbishment. Both of these factors shorten the life of the punch.

Before undertaking the replication of Clovis blades, Goode had spent a lot of time replicating large, thin bifacial tools using soft-hammer direct percussion—"soft-hammer bifaces." Such bifaces are common in prehistory, as are the distinctive soft-hammer bifacial thinning flakes produced in their manufacture. When Goode examined Clovis blades from the Keven Davis cache and casts of those found by Green (1963) at the Clovis site, his initial reaction was that these were a sort of specialized biface thinning flake that he could probably replicate fairly easily using direct soft-hammer percussion. His experimentation quickly bore this out. Goode's predilection for soft-hammer percussion was a strong factor in this, but he was also reacting to a widely held view that such blades could only be made by indirect percussion. Had someone else who was well versed in indirect percussion seen the same prehistoric examples and then began experimental replication, the result likely would have been equally successful replication in punched blades. With this in mind, it is nonetheless useful to consider some retrospective and general observations on his experience that Goode has made.

The problems a knapper faces in making blades are first to get suitable materials and knapping tools and then to prepare a core with correctly configured face and platform; after that the most difficult thing is accurate control of the point and angle of force. In the selection or preparation of the core face, the longitudinal curvature of the guiding ridge or crest is tolerant of less variability than one might assume. Too little curvature invites the initial blade to terminate in a hinge, whereas too much curvature is likely to result in the blade plunging and removing the base of the core. The platform and its angle to the face have to conform to even closer tolerances. If the raw material permits, the best chance of properly configuring a core comes when it can be shaped with flaking from at least three sides.

Finally, on the subjective consideration of skill, Goode notes that the several reasonably skilled modern knappers of his acquaintance either have not mastered or have not attempted blade making to the level of control reflected in Clovis blades. He feels that some could do it and that some could not. Of course no one knows what the ratio of success to failure would be among members of a society dependent upon stone tools, where knapping undoubtedly began at an early age, motivation was high, and a fully functional process of enculturating knappers was in place. Even under those conditions, it seems unlikely that everyone who tried could become an accomplished knapper or that every otherwise accomplished knapper could be a successful maker of blades. The implication of this view is that there was at least some degree of specialization among Clovis knappers; it is even possible that blade making was the domain of specialists or semispecialists who produced blades in exchange for goods or services.

BLADELIKE FLAKES

There are two final aspects of studying blades that require mention, both involving flakes that resemble blades. First are elongate, unretouched flakes that tend to meet Bordes's (1961) simple definition of blades; second are ordinary flakes that have been retouched to blade proportions. Neither of these forms is a blade, as we have just discussed, but each must be considered later in the discussion of Clovis lithic technology and in assessing whether the morphology of certain retouched pieces from Alabama and from Minnesota might be true blades.

Ordinarily I do not favor the use of equivocal terms such as *bladelike flake*, but in the case of Clovis technology this term can appropriately be applied to a consistently occurring, large form of flake that is relatively straight, has one or more dorsal ridges, and is often but not always twice as long as it is wide (Fig. 2.11). Flakes of this type are detached from large cores by direct percussion and do have a mix of flake and blade characteristics, hence offering some justification for the term *bladelike flakes*. At several Clovis sites there is evidence that Clovis points were fashioned from flakes of this kind.

It is necessary to address an issue related to blade recognition and to variability in blade assemblages. An intriguing aspect of blades from known Clovis contexts is their general lack of extensive invasive retouch. As the information to follow will show, there are some extensively retouched Clovis artifacts that may have been made on blades, but too few blade attributes remain for unequivocal diagnosis. There are assemblages, such as the Pelland blades from Minnesota and the

Figure 2.11. Large bladelike flakes from the Pavo Real site. Some preforms for Clovis points were made on such flakes, which have almost no longitudinal curvature.

Dust Cave artifacts from Alabama (discussed later), where invasive retouch is present on a majority of what appear to be blade tools. This poses a problem because it is relatively easy to laterally retouch certain kinds of flakes and produce objects almost indistinguishable from ones made on true blades. As an illustration of this, three experimentally produced bifacial thinning flakes (Fig. 2.12 a–c) were laterally trimmed to produce side scrapers of blade proportions (Fig. 2.12 d–f). Such objects as the large retouched piece from Blackwater Draw (J. Hester 1972: 106, 214, and Figs. 93 and 94) appear to be blade tools, but as easily could have been made on flakes. The best line of evidence for resolving this issue is comprehensive analysis of full assemblages where cores and debris from core reduction are available. This will not answer the question for each individual object but will afford the context for interpretation, since assemblages lacking unretouched blades, blade cores, and other debris from true blade production are not likely to contain retouched blades. Retouched bladelike artifacts found in settings lacking knapping debris, of course, remain problematical.

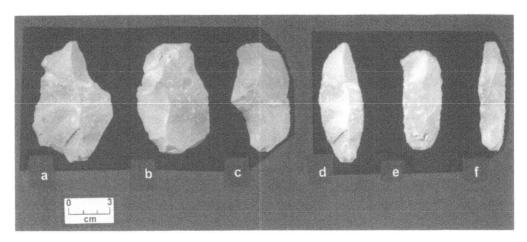

Figure 2.12. Three experimentally produced bifacial thinning flakes later trimmed to produce bladelike retouched pieces: (a, b, c) the bifacial thinning flakes; (d, e, f) the bladelike retouched pieces. Piece d is made from a; e, from b, and f, from c.

A Review of Clovis Lithic Technology THREE

ACROSS VIRTUALLY THE ENTIRE NORTH AMERICAN continent except for the area covered by the late Wisconsin-age ice sheet are found cultural remains that have been identified as "Clovis" (Fig. 3.1). Clovis manifestations, or local Clovis-related variants such as Gainey (Storck 1991), are recognized in unglaciated parts of Canada and the Great Lakes region of the United States (Irwin 1971; Keenlyside 1991; McDonald 1971) and in every quadrant of the United States' forty-eight contiguous states (Brennan 1982; Byers 1954; Dincauze 1993; Dunbar 1991; Frison 1991; Haynes 1966, 1982; Lepper and Meltzer 1991; Mason 1962; Morse and Morse 1983; Stanford 1991; Willey 1966; Willig 1991; Wormington 1957; but see Meltzer 1993). A few fluted points found in Alaska are attributed by some to Clovis (cf. Haynes 1966) but by others are considered historically to be only remotely related to Clovis (Clark 1991; Stanford 1991; Wormington 1957). These Alaskan fluted points may even be late derivatives of Clovis (Dixon 1993: 118–119). Fluted points called Clovis have been reported in Mexico (Ranere and Cooke 1991; Willey 1966) and Central America (Hester et al. 1982; Ranere and Cooke 1991; Willey 1966; Wormington 1957).

Archeologists place various artifact assemblages into the Clovis classification, but the principal diagnostic artifact is the lanceolate, fluted Clovis point (Krieger 1947; Sellards 1952; Wormington 1957). Other objects also are characteristic, including ivory points, large prismatic blades, and polyhedral blade cores (Haynes 1966; Frison 1991; Green 1963; J. Hester 1972; Stanford 1991). Clovis assemblages typically include stone artifacts made of high-quality materials, some of which are far removed from their geologic sources. Clovis artifacts are often found associated with remains of extinct fauna, and radiocarbon dating places most of them in the interval 11,200 to 10,900 uncalibrated radiocarbon years before present (Haynes 1992: 364). Unfortunately,

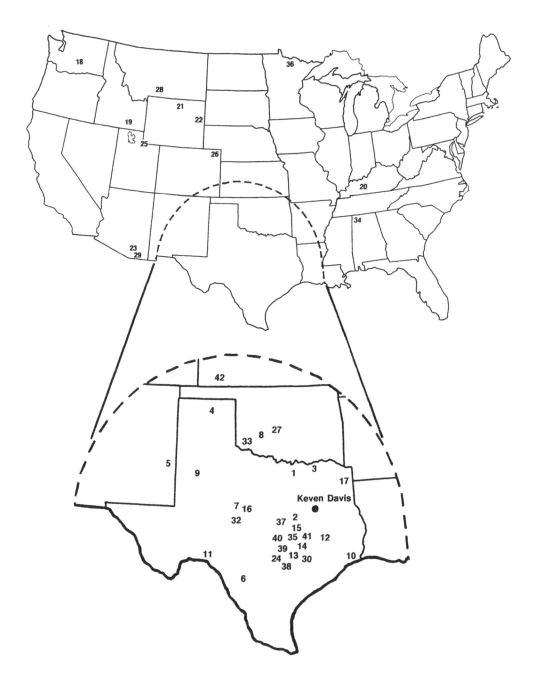

Figure 3.1. Location of sites with Clovis and other early cultural manifestations discussed in text. Numbers on map correspond to numbers in Table 3.1.

for many reported finds of Clovis points, the precise context and the nature of associated artifacts are unknown. In some cases this is because of the circumstance of discovery, in others it results from incomplete reporting.

In the literature, especially of the past two decades, "Clovis Culture" has come to be discussed in terms that closely conform to the archeological concept of horizon as defined by Willey and Phillips (1958: 33):

> *a primarily spatial continuity represented by cultural traits and assemblages whose nature and mode of occurrence permit the assumption of a broad and rapid spread. The archaeological units linked by a horizon are thus assumed to be approximately contemporaneous.*

Willey and Phillips relied to a large extent on stylistic markers to illustrate their concept but emphasized that horizons can be identified on the basis of "highly specialized artifact types, widely traded objects, new technologies" or "any kind of archaeological evidence that indicates a rapid spread of new ideas over a wide geographic space" (Willey and Phillips 1958: 32). In this sense, the concept would apply to much of what has been written about Clovis; in fact Willey has sometimes referred to Clovis as a horizon (Willey 1966: 63; Willey and Sabloff 1980: 161).

Explicitly or implicitly, other writers have characterized Clovis in a manner consistent with the Willey and Phillips concept. Consider, for example, the following definitions, descriptions, and characterizations of Clovis as a concept and as a "culture."

> *Clovis [as a concept] began in the 1930s as a local, western component, but by the 1950s stretched coast to coast as the first (and perhaps only) truly continental archaeological horizon. (Meltzer 1993: 294)*

> *The Clovis Cultural Complex is widespread over North America south of the area covered by the late Wisconsin ice sheets. Clovis artifacts demonstrate a remarkable similarity over the entire area. (Frison 1991: 321)*

> *Clovis archaeological culture . . . is widely believed to be the earliest phase of settlement in the New World. (Gramly 1993: 5)*

> *The technology of this [Clovis] period was apparently highly adaptive, allowing for the exploitation of a very large and diverse geographic area during a period of extreme climatic change and biotic reorganization. It was relatively short lived, lasting perhaps only 300 years. If Clovis peoples were the first migrants to the New World, the founding population must have been extremely large,*

highly prolific, and/or exceedingly mobile in order to populate the entire North American continent in such a short time. (Stanford 1991: 10)

While we await the outcome of this line of inquiry and because we are presently so ignorant of the true nature of Clovis lifeways, reluctantly and without advocating adoption of the usage, the remainder of this discussion will treat the stylistic and technological complex "Clovis" as an archaeological horizon, sensu Willey and Phillips. This begs the important question of what Clovis lifeways were like, but it provides a convenient rubric.

However, as more is learned, or as more assemblages are deemed to be "Clovis," it becomes more apparent that there is not a single constellation of traits that defines Clovis (Frison 1991; Meltzer 1993; Stanford 1991). Blades and tools made on blades, for example, are more characteristic of Clovis assemblages in some regions than in others. The uneven distribution of bone implements in Clovis assemblages is not entirely explainable by vagaries of preservation.

Horizon, as a concept, implies more homogeneity in cultural content than is found over the spatial expanse of Clovis manifestations. In effect, the presence of Clovis points, themselves quite variable, with or without any of the other "Clovis traits" is usually sufficient for classification of an assemblage as Clovis. Furthermore, each new Clovis discovery, almost without exception, expands the variability of the population of Clovis points. Thus, we are defining a horizon primarily on the basis of a single, variable artifact type.

The social and cultural realities behind the archeological manifestation of "horizon" are elusive in any case, but especially so in the Clovis case. We learn rather little about human adaptations by knowing that Clovis points are found so widely. Though perhaps descriptive, the horizon concept does not offer any explanation for the nature of the Clovis manifestation, nor does its closest rival concept, the "technocomplex."

Clarke (1968: 328–337) has suggested that diverse ethnic groups whose adaptation is basically similar may develop similar material cultures. These he refers to as "technocomplexes," defined as:

a group of cultures characterized by assemblages sharing a polythetic range but differing specific types of the same general families of artefact-types, shared as a widely diffused and interlinked response to common factors in environment, economy and technology. (Clarke 1968: 495)

In effect, *horizon* would imply that similarities in material culture derived from migration, diffusion, or stimulus diffusion of ideas associated with at least some degree of common ethnicity, whereas *technocomplex* could imply that material cultural similarities result more from similar adaptations than from shared history. The quotations cited above clearly connote shared origins among the diverse manifestations called Clovis and, therefore, lean closer to the horizon, than to the technocomplex, rubric. In my view, this leaning is justified on technological grounds.

Finished Clovis points could be similar wherever they are found yet be manufactured through the use of different knapping techniques. Wherever the evidence is available for early stages in the manufacture of Clovis points, remarkably similar knapping behavior is indicated. The same is true of Clovis blades. To me, this implies more than a small degree of historical relatedness. Furthermore, the concept of a technocomplex implies that similar adaptations are responsible for similar archeological assemblages, an argument that would be difficult to defend given the wide array of habitats exploited by the makers of Clovis points.

Meltzer (1993) has articulated, quite correctly, I think, a growing dissatisfaction with the notion that there existed across most of North America a single Clovis lifeway based on the hunting of large animals. As he observes, empirical evidence for such an adaptation is lacking, it is not logical to expect that such an adaptation prevailed in the diverse ecological settings where Clovis materials are found, and nothing can be found in the ethnographic record to support such an inference by analogy. Meltzer goes on to suggest that the historical accident that most of the early discoveries of Clovis artifacts were at sites where the big fossilized bones of large animals first attracted attention led to an interpretation of big-game hunting specialization. This interpretation developed such an inertia of its own, to continue Meltzer's argument, that it persists even in the face of a growing body of evidence to the contrary.

A few prehistorians are so steeped in the big-game hunting image of Clovis, that it has become the premise on which interpretations of the social makeup of Clovis hunting groups are based. In its most extreme form, this line of reasoning has led to the proposal that Clovis hunters were so well armed and hunted in such well-organized groups that they engaged in the "herd-confrontation" mode of attacking mammoths (see, for example, Saunders 1992). The notion derives from the characteristic defensive behavior of proboscidean herds, especially nursery herds composed mostly of adult females with their calves. When alarmed, such herds bunch in a tight circle for mutual defense with the juveniles in the center and the adults around the perimeter, virtually shoulder to shoulder, facing any would-be attacker. Should any member of the herd fall, the survivors close ranks and continue to stand their ground. Advocates of the herd confrontation model suggest that in this stance, the adult mammoths become ready prey that can be systematically killed off, one by one (e.g., Saunders 1992: 132). This line of reasoning absolutely fails to consider why rational hunters would confront precisely the mammoth social unit that evolved for defense and what these hunters could possibly do with multiple carcasses, each possibly weighing up to seven metric tons (Shipman 1992).

In short, a major rethinking of the romantic notion that Clovis big-game hunters ranged over most of North America is long overdue. It is not clear exactly what will replace that concept, and it will take years of good, hard dirt archeology to ascertain a clearer picture of Clovis "culture." Most likely, multiple regional Clovis adaptations will emerge, each with distinct responses to specific kinds of plants, animals, and other resources. It is to be hoped that there will also emerge some answers to the questions of why, over such a wide area, Clovis stone and bone tools were made in much the same way and why, in most regions, much use was made of exotic stone.

Defining this horizon is the Clovis point and various combinations of prismatic blades, blade cores, and points or foreshafts of bone or ivory. Use of ochre, various unifacial tools made on blades or flakes, as well as various bone tools is also indicated. Importantly, the technology of production of both the lithic and the bone or ivory artifacts is strongly similar over the wide geographic spread of this horizon. If similar response to similar adaptive needs—Clarke's technocomplex—were all that brought about the similarities in Clovis assemblages, only the finished products would be expected to be alike, not the details of their manufacture.

The best estimate of the temporal placement of this horizon is taken from Haynes (1992) as being 11,200 to 10,900 B.P. In addition to mammoth, remains of other large animals (such as bison and horse) as well as various small animals (including turtles, rabbits, and raccoons) are found in Clovis sites. Kill sites, campsites, caches, and burials are all known from this horizon, but isolated finds of Clovis points constitute the most common kind of occurrence (Dincauze 1993; Meltzer 1986, 1987, 1989; Story 1990).

The challenge facing archeology is to better understand in adaptive and behavioral terms the many and diverse local manifestations that constituted the relatively short-lived Clovis horizon, an effort far beyond the scope of this study. The purpose here is to bring clearer focus to blades and blade technology in Clovis times and to suggest some implications that this blade technology may have for more comprehensive interpretations.

In spite of repeated statements to the effect that Clovis is the earliest clearly recognized "culture," "complex," or "people" in the New
World (e.g., Haynes 1993: 219; Hofman et al. 1989: 29; Story 1990:
178; Willig 1991: 92;), which imply unanimity among archeologists as
to what "Clovis" is, the simple fact is, no consensus definition, concept, or interpretation of Clovis exists. The origin, distribution, dating,
material culture traits, economy, technology, environmental context,
and cultural-historical relationships to earlier, contemporary, or later
"cultures" are viewed differently by different scholars. In this study,
data from numerous Clovis sites are considered (see Table 3.1 for list
and for references).

TABLE 3.1. Clovis and other early sites and components considered in this study.

Map No.	Site, Location, Character	References
1	LEWISVILLE; north-central Tex.; open campsite	Crook and Harris 1957; Holliday 1997; Stanford 1982, 1983; Story 1990
2	HORN SHELTER 2; central Tex.; rockshelter campsite	Holliday 1997; Redder 1985; Story 1990; Watt 1978
3	AUBREY; north-central Tex.; open campsite	Ferring 1989, 1990, 1994; Holliday 1997; Humphrey and Ferring 1994
4	MIAMI; Tex. Panhandle; open kill site	Holliday 1997; Holliday et al. 1994; Sellards 1938, 1952
5	BLACKWATER DRAW; east-central N.M.; open camp, kill, cache site	Evans 1951; Green 1963; J. Hester 1972; Holliday 1997
6	KINCAID; central Tex.; rock-shelter campsite	Collins 1990a; Collins et al. 1989; Holliday 1997
7	McLEAN; west-central Tex.; open kill site	Holliday 1997; Ray and Bryan 1938; Sellards 1952
8	DOMEBO; west-central Okla.; open kill site	Holliday 1997; Leonhardy 1966
9	LUBBOCK LAKE; Tex. Panhandle; open kill or scavenge and butchery site	Holliday 1997; Johnson 1987, 1991
10	McFADDIN BEACH; southeastern Tex.; secondary deposit	Long 1977; Turner and Tanner 1994; TARL files
11	BONFIRE SHELTER; Lower Pecos, Tex.; rockshelter camp? site	Bement 1986; Dibble and Lorrain 1968; Holliday 1997

12	DUEWALL-NEWBERRY; east-central Tex.; open kill or scavenge site	Carlson and Steele 1992; Steele and Carlson 1989
13	SPRING LAKE; central Tex.; secondary deposits	Shiner 1981, 1982, 1983; Takac 1991
14	VARA DANIEL; central Tex.; open camp? site	Collins et al. 1990; Ricklis, Blum, and Collins 1991
15	GAULT; central Tex.; open camp, ritual? site	Collins et al. 1991; Collins, Hester, and Headrick 1992; Hester, Collins, and Headrick 1992
16	YELLOW HAWK; west-central Tex.; open quarry/workshop	Mallouf 1989
17	MURPHEY; northeastern Tex.; open kill? site	Story 1990
18	EAST WENATCHEE; central Wash., open cache site	Gramly 1993; Mehringer 1988, 1989; Mehringer and Foit 1990
19	SIMON; south-central Idaho; open cache site	Butler 1963; Woods and Titmus 1985
20	ADAMS; western Ky.; open workshop	Sanders 1990
21	COLBY; north-central Wyo.; open kill site	Frison and Todd 1986
22	SHEAMAN; east-central Wyo.; open kill site	Frison 1982
23	MURRAY SPRINGS; southeastern Ariz.; open kill site	Haynes 1993; Hemmings 1970
24	PAVO REAL; central Tex.; open camp, workshop	Henderson and Goode 1991
25	FENN; probably in or near southwestern Wyo.; cache site	Frison 1991
26	DRAKE; northeastern Colo.; open cache site	Stanford and Jodry 1988
27	ANADARKO (McKEE); central Okla.; open cache site	Hammatt 1970
28	ANZICK; south-central Mont.; shelter? burial/cache site	Lahren and Bonnichsen 1974; Wilke, Flenniken, Ozbun 1991
29	LEHNER; southeastern Ariz.; open kill site	Haury, Sayles and Wasley 1959
30	unnamed; east-central Tex.; open cache? site	C.K. Chandler, personal communication 1994

31	KEVEN DAVIS; northeastern Tex.; open cache site	Collins 1996; Young and Collins 1989
32	41RN108; west-central Tex.; open camp? site	Bryan and Collins 1988
33	CEDAR CREEK; southwestern Okla.; open cache? site	Hammatt 1969
34	DUST CAVE; northwestern Ala.; shelter campsite	Driskell 1994
35	WILSON-LEONARD; central Tex.; open campsite	Collins et al. 1993; Holliday 1997; Masson and Collins 1995
36	PELLAND; northern Mont.; open cache? site	Stoltman 1971
37	EVANT; central Tex.; open cache? workshop	Goode and Mallouf 1991
38	COMANCHE HILL; central Tex.; open cache? site	Collins and Headrick 1992; Kelly 1992
39	EISENHAUER; central Tex.; open cache? workshop	Chandler 1992
40	41LL3; central Tex.; open camp? site	Collins 1990b; TARL files
41	CROCKETT GARDENS; central Tex.; open cache? site	McCormick 1982
42	SAILOR-HELTON; southwestern Kans.; open cache	Mallouf 1994

Regional variants are emerging from recent work (cf. Dunbar 1991; Morse and Morse 1983; Shott 1993; Stanford 1991; Willig 1988) and regionally distinct adaptations have been suggested; even the hallmark of mammoth-hunting specifically or big-game hunting generally is being ever more closely scrutinized (Collins and Kerr 1993; Dincauze 1993; Ferring 1989; McDonald 1971; Meltzer 1993). The nature of Clovis subsistence for the southern plains and plains periphery is pertinent to any effort to interpret the Keven Davis blade cache.

Several sites on and near the southern plains have contributed evidence on Clovis subsistence. These are Lewisville, Horn Shelter 2, Aubrey, Miami, Blackwater Draw, Kincaid, McLean, Domebo, Lubbock Lake, and possibly McFaddin Beach. Bonfire Shelter and Duewall-Newberry lack diagnostic Clovis artifacts but are possibly affiliated kill, processing, or scavenging localities.

Some important generalities emerge from the evidence at these several sites. Clovis pyrotechnology evidently did not include the use of stone as a heating element, a surprising fact given its long antecedence in parts of the Old World (de Sonneville-Bordes 1989; Movius 1968; Perles 1976). Habitation sites occupied in Clovis times included rock-shelters (Kincaid, Horn, and possibly Bonfire) and open localities. Kincaid is noteworthy not only as an occupied rockshelter but as one in which Clovis peoples invested great effort in hauling river boulders and cobbles into the shelter and paving a portion of the floor. Open sites occur primarily along streams or near springs (Lewisville, Spring Lake, Vara Daniel, Gault, and Aubrey). Knapping transpired at habitation sites near abundant sources of high-quality chert (Kincaid, Gault), at habitation sites remote from chert sources (Aubrey), and at quarry localities apart from habitations (Yellow Hawk). Some Clovis habitation sites, such as Horn Shelter 2 and Lewisville, evidently saw little knapping other than the removal of relatively few, very small (resharpening?) flakes.

Clovis lithic caches are indicated, with varying degrees of certainty, at Blackwater Draw in New Mexico and Sailor-Helton in Kansas, as well as at several sites in Texas, namely Keven Davis, Evant, Comanche Hill, McFaddin Beach, Crockett Gardens, and an unnamed site in Bastrop County. These and the better-known Clovis caches—Fenn, Anzick, East Wenatchee, Drake, and Simon—are discussed in more detail later.

Carcasses of large animals (notably mammoth, bison, and perhaps mastodon, some of which may have been hunted and killed, others of which may have died naturally or had a natural death hastened by humans) were processed at such sites as Aubrey, Domebo, Lubbock Lake, McLean, and Miami, as well as possibly Duewall-Newberry, McFaddin Beach, and Murphey. Data on plant foods are virtually nonexistent for Clovis, but a diverse faunal menu can be inferred from several sites: Aubrey—turtle, bison, deer, and small mammals; Blackwater Draw—mammoth, horse, camel, and bison; Horn Shelter 2—bison, large land turtle, aquatic turtles, fish, small mammals, and possibly land snails; Kincaid—mammoth, horse, box turtle, badger, raccoon, alligator, and slider turtle; Lewisville—box turtle, deer, rabbit, raccoon, bird eggs, river mussels, snake, fish, frog, lizard, and perhaps snails; and Lubbock Lake—mammoth, horse, camel, bison, giant bear, and giant armadillo.

Clovis peoples on the southern periphery of the Great Plains were not highly specialized, migratory mammoth hunters. They were fairly generalized hunter-gatherers exploiting a wide array of animal, and probably plant, resources. This is not to say that they were not successful hunters of mammoths and other large game. Their sophisticated weaponry and apparent kill sites attest to their hunting capabilities. As an aside, it is worth noting that bone or ivory foreshafts are commonly inferred to be part of compound darts or spears used by Clovis hunters. If this were the case, some evidence of these should have been found in kill sites where points were found with preserved animal bone. Either Clovis hunters were extremely proficient at retrieving the foreshafts and points, or foreshafts were not in common use. The technology of stone-tool production and caching of lithic artifacts are consistent with this interpretation of Clovis subsistence.

CLOVIS LITHIC TECHNOLOGY

To understand Clovis blades from the perspective of production technology, it is helpful to consider the entire range of Clovis knapping behavior. No single site that provides evidence on the full spectrum of Clovis knapping—with the possible exception of the Adams site in Kentucky—has yet been reported, but a fairly complete account can be pieced together from complementary bits of incomplete evidence at a number of sites. Data have been gleaned from sites across much of North America (including Adams, Blackwater Draw, Colby, East Wenatchee, Sheaman, Simon, and Murray Springs), but sites in the south-central United States, because they were more readily accessible to me, have provided most of the examples discussed and illustrated (Blackwater Draw, Eisenhauer, Gault, Kincaid, Pavo Real, site 41LL3, Vara Daniel, Wilson-Leonard, and Yellow Hawk), which undoubtedly introduces a regional bias into this account.

Two major reductive strategies have been recognized for the production of chipped-stone artifacts of Clovis affinity—bifacial reduction and prismatic blade production (Collins 1990b; Sanders 1990), but the distinction is not absolute. There is evidence to suggest that large bladelike flakes were often preferred for the initiation of biface production. These bladelike flakes are also a characteristic form of debitage rendered by the initial steps of prismatic blade production. Clovis implements are made on bifaces, blades, and flakes. Most of the flakes selected for modification into tools seem to be the byproducts of biface and blade-core work rather than products of separate flake-core reduction, and, therefore, are not here considered to be part of a distinct reductive strategy.

CLOVIS BIFACES

Biface production is ubiquitous and remarkably uniform over the continent in Clovis times, whereas blade production, though equally uniform, is less common overall and virtually absent, or at least not recognized or not reported, in some regions. Projectile points seem to be the usual end product of bifacial reduction, but other forms of bifaces are sometimes found (Stanford 1991).

Bifaces were fashioned from large bladelike flakes (Figs. 2.11 and 3.2 a) or cores in most cases (Bradley 1982, 1993; Callahan 1979; Mallouf 1989; Morrow 1995; Sanders 1990), although, locally, constraints imposed by the nature of raw materials affect the technology (cf. McDonald 1968). Most such bifaces seem to have been Clovis point preforms rather than finished implements and typically are of high-quality materials. Direct percussion, probably hard-hammer in initial reduction followed by soft-hammer in thinning and most trimming, was used in all but final edge trimming, where some pressure-flaking may have been employed. Early-stage bifaces (Fig. 3.2 b) were fashioned with minimal platform preparation and relatively few flake removals. Platforms were produced by roughly chipping a bevel along the edge, with platform grinding used increasingly as flaking progressed. Broad flakes extending completely or nearly across the width of the biface were removed, and overshot flakes occur frequently. Once achieved, a generally lanceolate outline with straight base, convergent tip, and straight to slightly convex edges was maintained throughout the flaking process (Fig. 3.3 a, b). The base was beveled with a few flake removals to serve as the first fluting platform (Fig. 3.3 c, d). Subsequently, the opposite face was beveled to accommodate the second fluting (Fig. 3.4 a). Flutes appear to have been produced by direct percussion, or, rarely, by indirect percussion. Final trimming, employing both percussion- and pressure-flaking techniques, resulted in even edges and symmetrical outlines centered on the flutes (Fig. 3.4 b). Some final flaking intrudes on the flute scars, and sometimes a flake was directed from the lateral margin to remove any hinge or step fracture at the distal terminus of the flute, although this characteristic is not as common as implied by Bradley (1982, 1993). Finally, lateral edges were ground. Complete, pristine points generally exceed 100 mm in length. Points were repeatedly resharpened until discarded at 50 mm or less in length (Fig. 3.4 d, e). The best evidence for this discard behavior comes from the Kincaid and Yellow Hawk sites, where unbroken points nearly 50 mm long were found along with the debris from bifacial reduction (Fig. 3.4 d, e). In both of these cases, the discarded points are of the same local chert as the flaking debris, indicating return to the same chert source within the use-life-span of the points.

The two shortest (ca. 40 mm in length) of the full-sized Clovis points examined in this study (Fig. 3.4 f, g) are from sites on the southern half of the Llano Estacado of western Texas, where chippable stone is not available. It would appear that these were used longer and resharpened more than points found in areas of abundant chert of high quality.

Short versions of full-sized Clovis points resulting from repeated

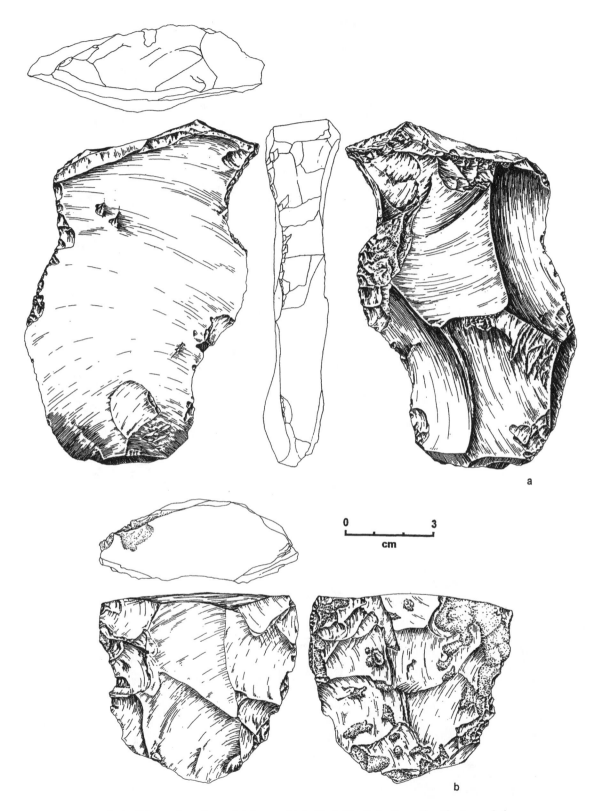

Figure 3.2. Clovis flake and preform: (a) large, thick, bladelike flake from the Clovis workshop site of Yellow Hawk; and (b) nearly plano-convex, early-stage bifacial preform, evidently fashioned from a large flake, from the Clovis component at Kincaid Rockshelter.

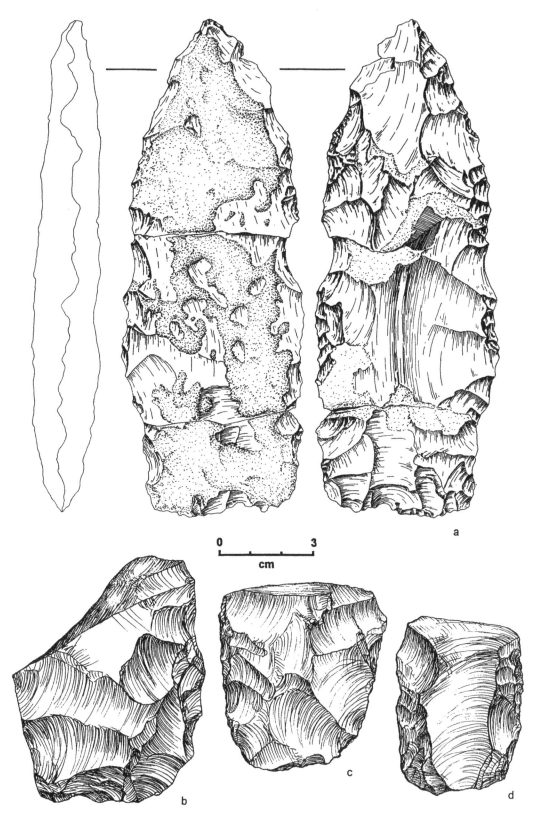

0   3
cm

a

b

c

d

Figure 3.3. Clovis bifacial preforms: (a) early-stage preform from Vara Daniel site; and (b, c, d) sequence of three preforms from Kincaid Rockshelter, the most advanced of which was ruined by the first fluting attempt.

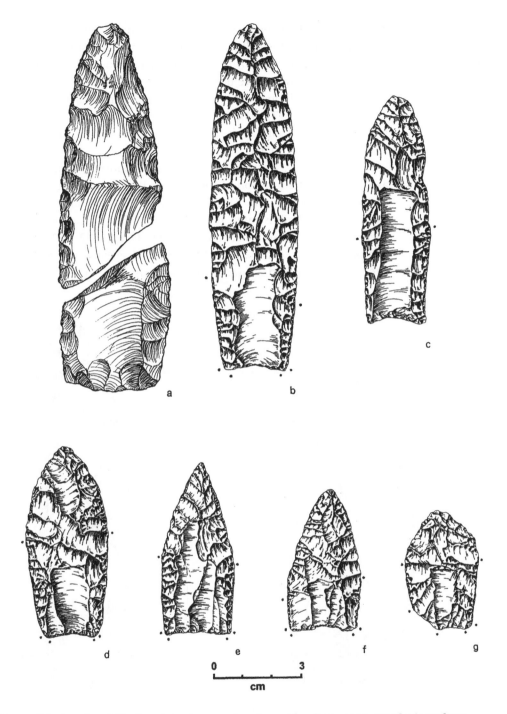

Figure 3.4. A series of Clovis points at successive stages of resharpening: (a) Clovis preform, from Kincaid Rockshelter, ruined by perverse fracture after successful removal of the first flute. (b) complete point with little or no resharpening, from the Miami site; (c), resharpened point, from the Miami Site; (d, e) two complete, extensively resharpened points abandoned at workshops (d from Yellow Hawk, e from Kincaid); and (f, g) two extremely resharpened points, from the chert-barren Llano Estacado of Texas (f from Lubbock Lake, g from Midland County).

resharpening of longer ones are not to be confused with another Clovis artifact form—very small points manufactured at a diminutive scale (J. Hester 1972: 97–102). Although these seem too small to be effective as weapon tips, they have been found in association with mammoth carcasses in positions to suggest they were part of a lethal arsenal (J. Hester 1972: 97–102). Fluting and the closely related issue of hafting of full-sized Clovis points are of interest to anyone who studies these specimens.

Thicker bifaces, less refined than point preforms, occur in Clovis assemblages. The function of these is not known, and their numbers seem to be relatively low. One from Yellow Hawk (Mallouf 1989: 93–94, Fig. 10 a) is illustrated here (Fig. 3.5). Other examples include specimen EL114 from Blackwater Draw (J. Hester 1972: 219, Fig. 90 k) and probably the several said to be from the deepest levels at the Gault site (David Olmstead, personal communication 1991).

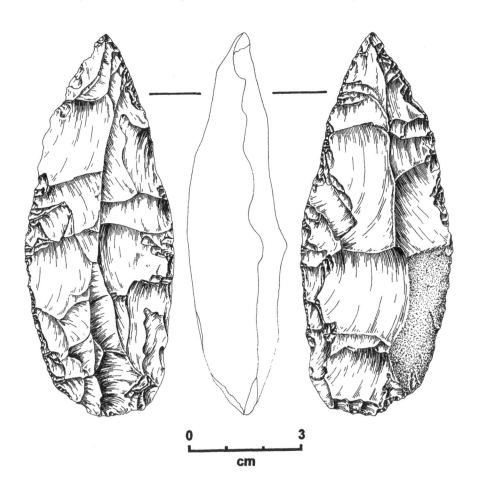

0   3

cm

Figure 3.5. Thick Clovis biface from Yellow Hawk. This form of biface is not a point preform.

Prismatic blades were struck from large prepared cores of two types, conical and wedge-shaped. By far the more common cores are conical with the general plane of the platform at right angles to the long axis of the core and to the proximal blade facets (Figs. 3.6, 3.7, 3.8, 3.9, 3.10, and 3.11). After removals, multiple blade facets form a convex face, sometimes around the full circumference of the core. Although the platform plane is approximately perpendicular to the proximal core face, it is composed of multiple, short, deep flake scars emanating from its periphery (Figs. 3.6 a–d and 3.8 a–d). The negative bulbar scar of each of these flakes produces an acute angle of approximately 60 to 70 degrees with the core face. It is inferred that, in many cases, a punch was placed in these concavities for blade removal as the ca. 60- to 70-degree platform angle is retained on the blades (or, since the complementary angle is measured, 110 to 120 degrees). In some cases, the platform maintenance flakes are positioned to create a promontory at the point where a ridge on the core face intersects the core platform; these would almost certainly be set up for direct percussion. Platform maintenance flakes commonly terminate in hinges with the cumulative effect of producing large central knots on the platforms (Figs. 3.6 a–d and 3.8 a, b, d). Such platforms were rejuvenated by removal of core-tablet flakes (Figs. 3.7 a–d, 3.8 c, and 3.12 a).

Wedge-shaped cores (Fig. 3.13) in comparison with conical cores have a different overall shape and, most significantly, have an acute angle between the platform and the core face. These cores generally have a relatively narrow face and the platform is usually multifaceted. Maintenance of platforms on these cores between blade detachments is much simpler than on the conical cores and consists of trimming an acute, bifacial edge. Wedge-shaped cores can easily accommodate two opposing platforms. If blades are detached from the same core face in alternating fashion from opposing platforms, the blades will be less curved in longitudinal section (Wilke and Quintero 1994).

Cores often retain cortex (Fig. 3.6 a), and some cortex is seen on blades (Fig. 3.14 a, b). The nature of the cortex, particularly on some blades (e.g., Fig. 3.15 a–h), reveals that rather than always preparing bifacial crests to direct the primary blade, Clovis knappers sometimes selected raw material with natural ridges and initiated core reduction along them. There are examples where the entire ridge is cortical (Figs. 3.14 a and 3.15 c, d) and others where stream-battering or some other natural spalling created a ridge that was used (Fig. 3.15 g). An end scraper from the Wilson-Leonard site was fashioned on a primary blade, the ridge of which was partly natural cortex and partly a bifacially prepared crest (Fig. 3.16 a). When a cortical nodule or cobble is reduced as a blade core (e.g., Fig. 3.6 a), several cortically backed blades may be produced in a series until the cortex is removed

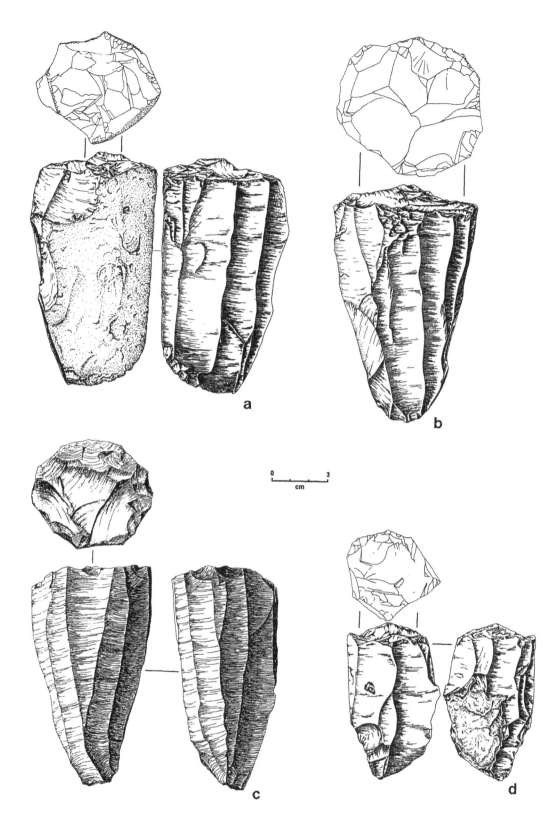

Figure 3.6. Conical Clovis blade cores with characteristic platforms: (a) from site 41LL3; (b) from the Gault site; (c) from Comanche Hill (Van Autry); and (d) from Kincaid Rockshelter. Note presence of knots on a and d. Note also that most blade scars lack negative bulbs as a result of platform removals.

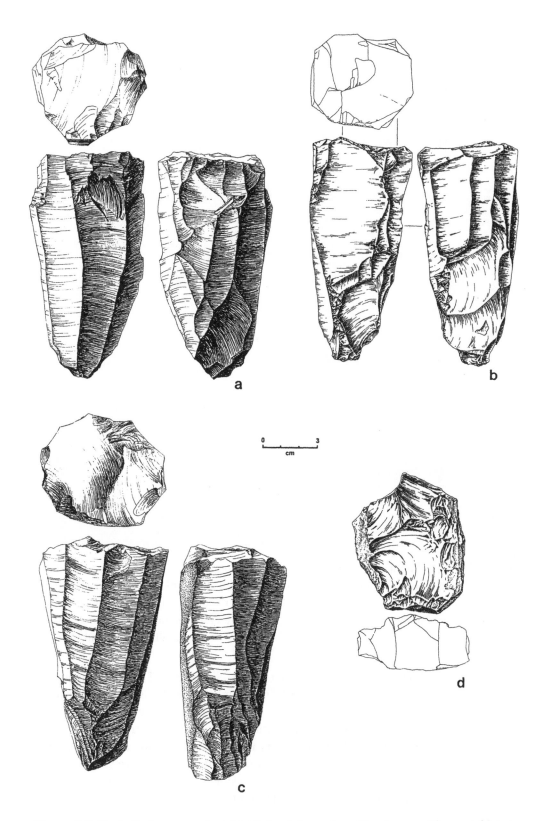

Figure 3.7. Clovis blade cores and a tablet flake: (a) core from Eisenhauer, with core-tablet scars; (b) core from site 41GL175, with core-tablet scars; (c) core from Comanche Hill, with core-tablet scars; and (d) a core-tablet flake from Pavo Real.

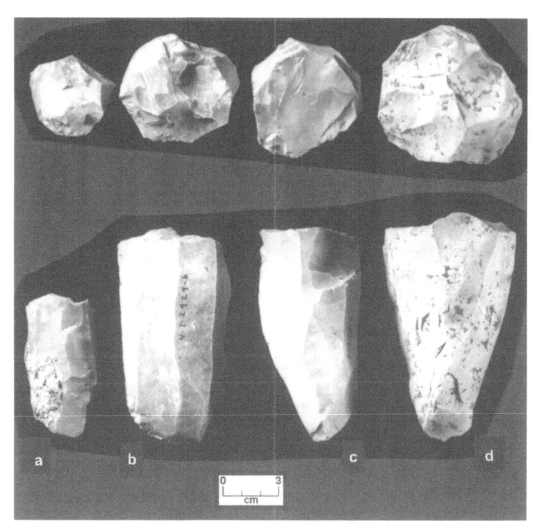

Figure 3.8. Typical conical Clovis blade cores with platforms in various stages: (a, b) platforms with large knots resulting from repeated removals of small flakes around perimeter of platform; (c) freshly rejuvenated platform with large scar from tablet removal; and (d) platform with several small flake removals, but no knot formed. Sites represented are (a) Kincaid, (b) site 41LL3, (c) Eisenhauer, and (d) Gault.

(Figs. 3.14 b and 3.15 h–j, l, m). Six stages of blade reduction are defined later in this report (Chapter 5) on the basis of blade attributes.

As a blade core is being reduced, the ridges left after each blade detachment are the potential guiding arrises for any subsequent detachment. Some secondary blades show crushing and flaking along the ridges on their exterior surfaces. In some cases (e.g., Figs. 3.15 n and 3.16 g), this may be the result of correcting irregularities that could have caused misdirection of the blade.

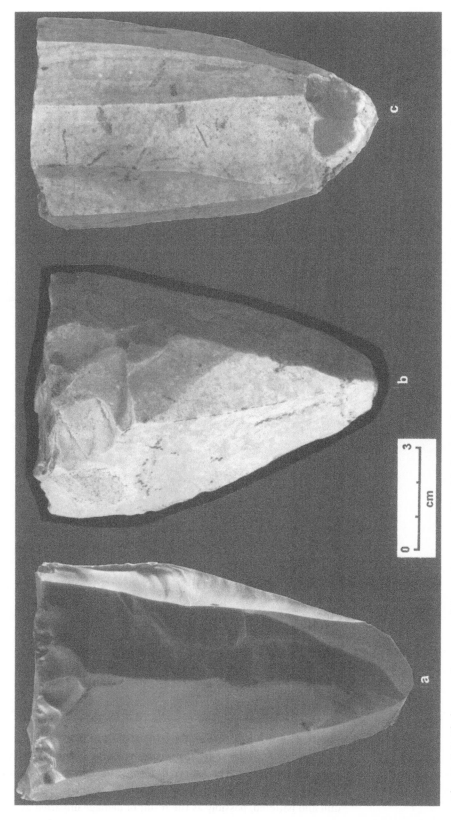

Figure 3.9. Blade cores from central Texas documented in old photographs in the records of the Texas Archeological Research Laboratory: (a) from Medina County; (b) from Travis County; and (c) from Williamson County. Note the heavy cortex on the Williamson County specimen revealed by plow damage near the core tip.

Figure 3.10. Blade core from the surface of site 41BX998 near San Antonio: The core has (a, c) two wider faces, and (b, d) two narrower edges. The pattern of flake scars indicates that most blade removals were from the two wider faces; the two edges are dull and battered, possibly as the result of damage imparted as the core was supported or held during knapping.

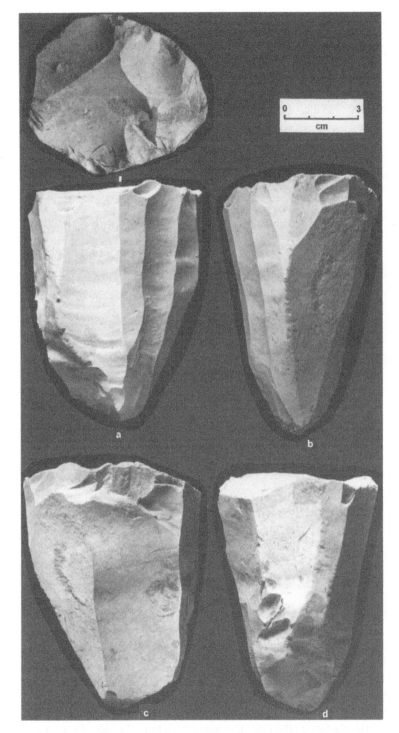

Figure 3.11. Blade core from the surface of the Gault site (41BL323).
Like the core from site 41BX998 in Figure 3.10, this specimen has
(a, c) two wider faces and (b, d) two narrower edges, with the pattern
of blade scars indicating most blade removals as being from the two
wider faces, and the two edges as being dull and battered, possibly
from damage imparted as the core was supported or held
during knapping.

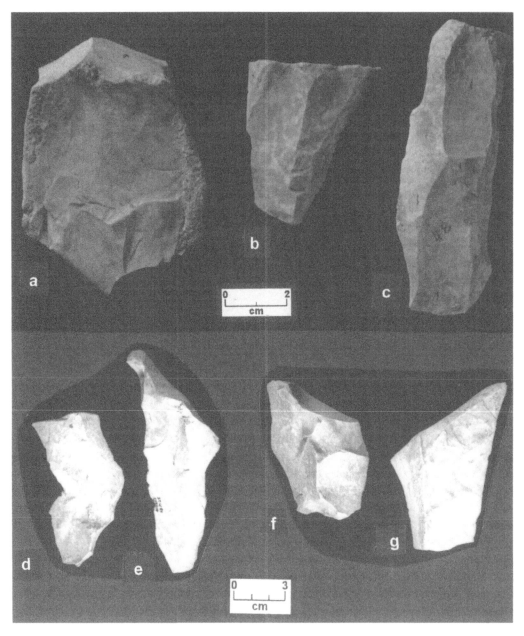

Figure 3.12. Various forms of core maintenance flakes: (a) core-tablet showing the platform that was removed; (b) hinge-recovery flake detached from core face along with a portion of the lip around the perimeter of the platform and two proximal blade scars, one of which terminated in a hinge; (c) hinge-recovery flake struck from the distal end of the core; (d) partial core-tablet flake showing the portion of the platform removed; (e) large flake that was struck from distal end of the core, removing a portion of the platform; (f) a flake that was struck from the side of the core, removing a portion of the core face (to the right) and a portion of the platform (to the left); (g) a flake that was struck from the side of the core, removing the distal end of the core (note distal terminations of blades on right side of the flake).

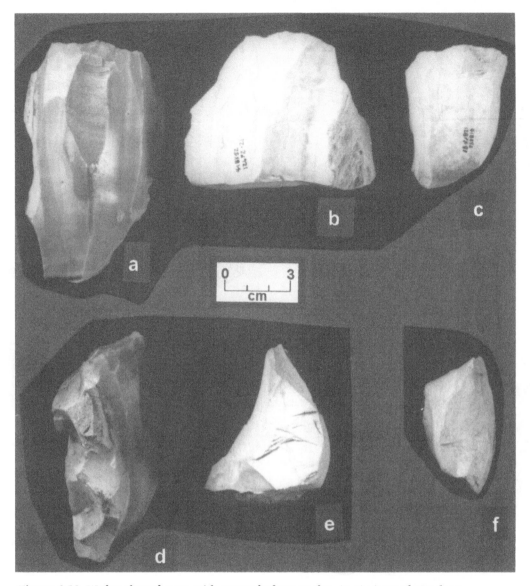

Figure 3.13. Wedge-shaped cores with acute platform angles: (a–c) views of core faces, and (d–f) edges, showing acute angle between core face and platform. Sites represented are (a, d) Gault, and (b, c, e, f) Pavo Real.

The very infrequent presence of negative bulbar scars on core faces (especially on conical cores) and on the exterior of blades indicates that platform rejuvenation (core tablet removal) was done at frequent intervals during the detachment of blades. Debris in Clovis workshops also indicates that blades sometimes terminated prematurely in hinges or step fractures (Figs. 3.7 a, b; 3.8 c; 3.15 d; and 3.17 h). Some of these were corrected with hinge recovery flakes or blades struck in

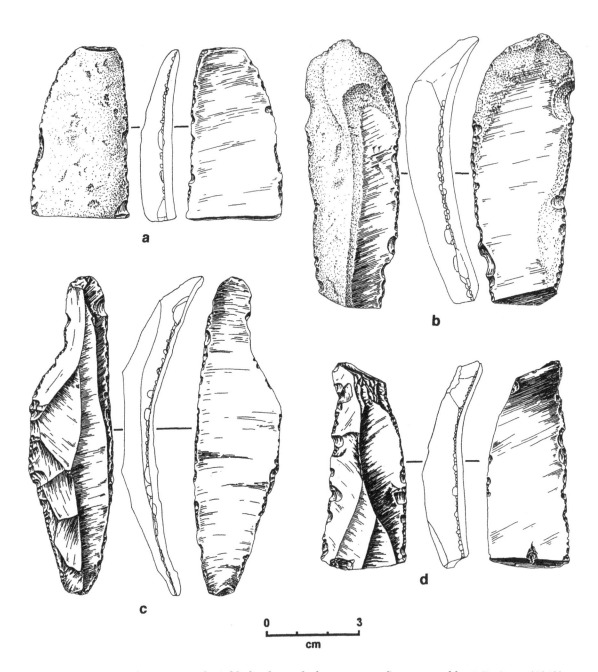

Figure 3.14. Clovis blades from Blackwater Draw first reported by F. E. Green (1963):
(a) primary cortex, (b) secondary cortex, (c, d) noncortex. (Drawn from casts in TARL collections.)

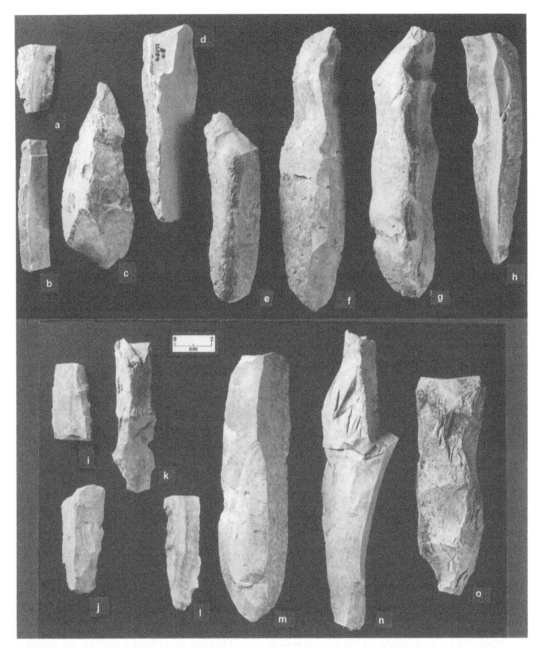

Figure 3.15. Typical early-stage Clovis blades from Pavo Real (see Chapter 4): (a–h) stage 1 with natural exterior surfaces, cortex or partially cortex; (i–o) stage 2 with prepared crests. Note that two of those with prepared crests are modified with (i) marginal retouch and (j) steep transverse flaking—an end scraper.

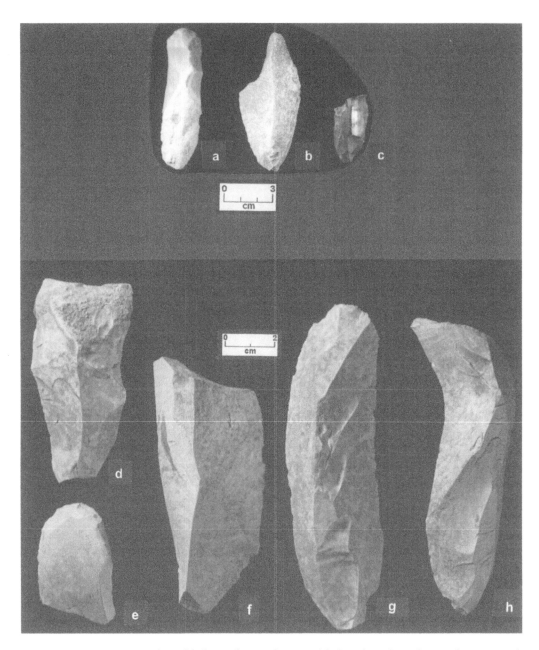

Figure 3.16. Tools made on blades and typical stage 3 blades: (b, c) lateral retouch; (a, e) end scrapers; (d, f, h) stage 3 blades with prior blade scars on exterior; and (g) blade with battering on the prominent dorsal ridge. Sites represented are (a–c) Wilson-Leonard and (d–h) Pavo Real.

the longitudinal axis of the core, either from the primary platform or the distal end of the core (Fig. 3.12 b, c). In extreme cases, a hinge recovery flake or other core-maintenance detachment might be removed transversely or diagonally across the face of the core from a platform along the margin of the core face (Fig. 3.12 d–g).

Blades often have minute platforms, almost no bulbs, minimal ripple marks on the interior surface, and strong curvature (Figs. 3.14 and 3.18 a). Light to moderate grinding appears on some blade platforms (Fig. 3.19). Some ridges on blade exteriors as well as on core faces show bruising, apparently in part from curation and possibly transport between episodes of blade removal. It is also possible that some of this is wear from use as an obtuse-angle planer (Crabtree 1973). Blades were utilized intact or segmented and retouched into end scrapers and other forms (Figs. 3.15 j; 3.16 a, b, e; 3.17 a, l, n, o; and 3.18 b–d). They were not made into points.

As Newcomer (1975), Whittaker (1994), and others have observed, blade attributes cannot be used definitively to distinguish direct soft-hammer percussion from indirect-percussion blade removals. However, the two forms of cores discussed here do bring further evidence to bear on the question. The more-common conical cores, because of the manner of platform maintenance typically utilized, quickly develop the protruding knot near the center of the top of the core. This knot is so prominent on some cores that, if direct percussion were attempted, it would be in the arching path of a percussor as it approached the optimal point of impact at a sufficiently low angle for successful blade detachment. Only indirect percussion could be used to accurately direct force at the proper point and at the correct angle in the sunken platform areas around the perimeter of the top of these conical cores. Other conical cores, however, have less-prominent knots, and here direct percussion is not only feasible, but perhaps likely: Goode's experimental percussion blades are frequently driven off of such cores (see Chapter 2).

The less-frequent wedge-shaped cores, on the other hand, are ideally suited for direct-percussion blade detachment. On these the platform is prominent and the pathway for proper delivery of a direct-percussion blow is clear of any obstruction.

The ridges on most blade cores exhibit discontinuous nicking and bruising (e.g., Fig. 3.20). In some cases the cause is probably damage subsequent to loss or disposal by the original knappers. The core (Fig. 3.7 a, 3.8 c) from the Eisenhauer site (41BX959 [Chandler 1992]) and the Evant Cores (Goode and Mallouf 1991), for example, have extensive nicking and bruising on their ridges. These cores also have rusty metallic streaks on several surfaces and, particularly, on the ridges. It is apparent that metal plows or discs hitting these pieces caused this

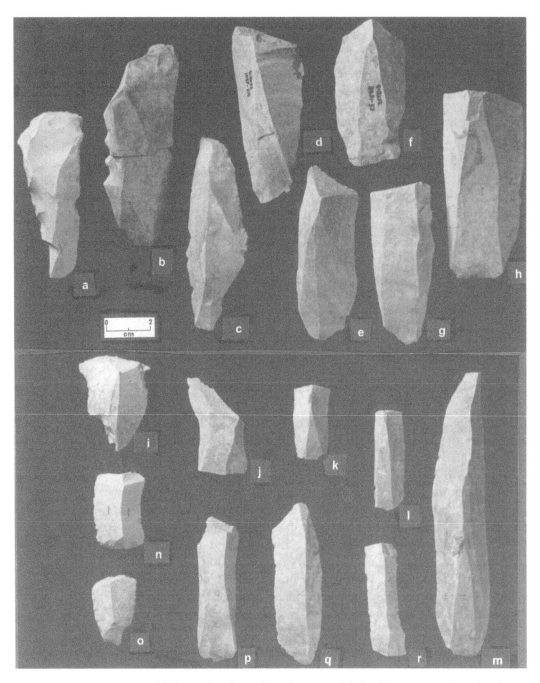

Figure 3.17. Late-stage blades and tools made on late-stage blades from Pavo Real: (c–h, j, k, p) stage 5 with regular edges, no or very little cortex, prior blade scars; (l, m, q, r) stage 6 with very regular edges and prior blade scars; (a, i) scrapers on stage 5 blades; (n, o) scrapers on stage 6 blades; (b) refitted pieces of intentionally segmented stage 5 blade that required multiple, sharp blows to the interior surface to bring about the break; what appears to be retouch along left margin is damage caused in attempting to segment this thick piece.

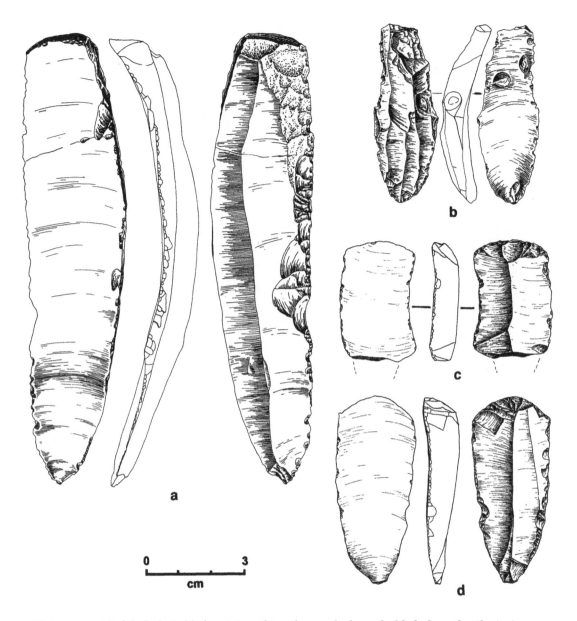

Figure 3.18. Modified Clovis blades: (a) machine-damaged edges of a blade from the Clovis site; (b) intentional edge trimming of a heat-damaged blade segment from the Gault site; and (c, d) end scrapers on blade segments from Pavo Real.

damage. In other cases, such as the core from 41LL3 (Figs. 3.6 a and 3.8 b) with similar damage (Fig. 3.20) and the Evant Cores (Goode and Mallouf 1991), the cause is unknown, as no metallic staining is present. However, the original contexts at the sites, relative to possible plow damage, are not known in detail. Bruising and nicking on the ridges of the core from the Gault site (Figs. 3.6 b and 3.8 d)

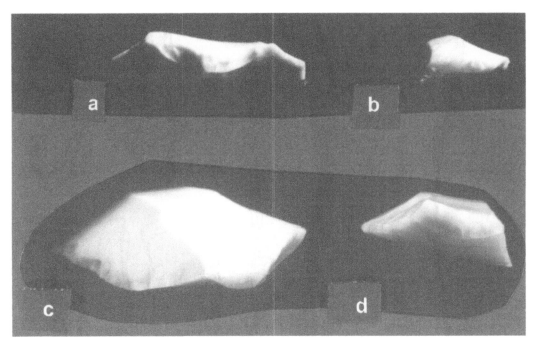

Figure 3.19. Clovis blade platforms showing light to moderate grinding for platform strengthening.

are definitely prehistoric in age, as this core was recovered in primary subsurface context in an area that has never been plowed. Possible explanations for this damage are discussed further in Chapter 9.

Clovis blades consistently have small bulbs and smooth interior surfaces, attributes inconsistent with hard-hammer percussion but equally consistent with punch techniques and with direct, soft-hammer percussion. The aggregate evidence, then, suggests that both techniques—direct soft hammer and indirect (punch)—were used, apparently with the indirect technique perhaps being the more common. The large core from the Gault site (Fig. 3.21) is of the more-common conical form, and all but one of its ten blade scars originate from the typical, broad conical-core platform. It is presumed that these nine were detached through use of the punch technique. A single blade scar originates from the conical end of the core from a minimally prepared, but highly acute, platform. This latter platform and the morphology of the blade scar (smooth, small negative bulb) suggest that direct, soft-hammer percussion was the technique used in its detachment.

Conical blade cores with the distinctive characteristics just described have been found at several localities in central Texas over the years. Three examples from old photographs in the records of the Texas Archeological Research Laboratory clearly exemplify these characteristics (Fig. 3.9). The cores are from Medina (Fig. 3.9 a), Travis

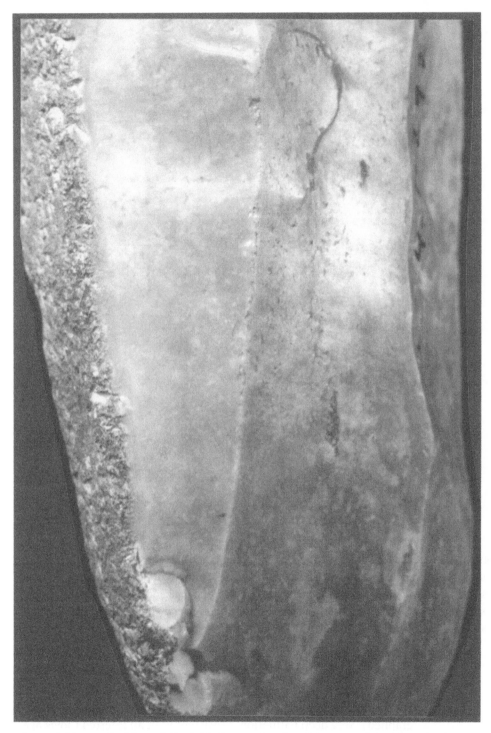

Figure 3.20. Detail of face of Clovis blade core showing chipping and bruising of ridge between two blade scars.

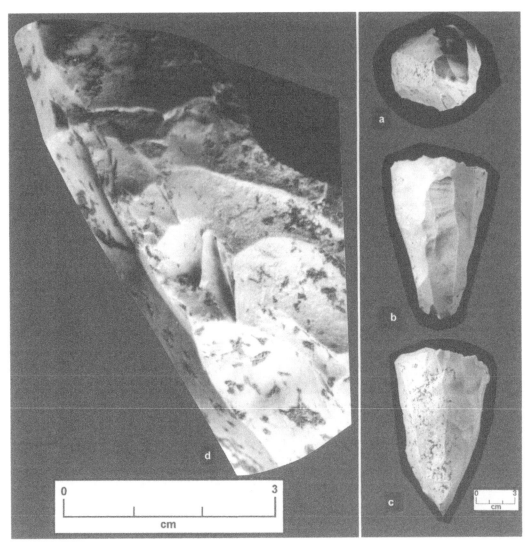

Figure 3.21. Conical blade core with secondary platform at distal end (the secondary platform forms an acute angle with the core face that is closely similar to those seen on wedge-shaped cores): (a) distal end of core showing acute platform and proximal end of blade scar; (b) face of core showing scar of blade detached from acute platform; (c) lateral view of acute platform with scar of blade detached therefrom on the right; and (d) detail of preparation of acute platform.

(Fig. 3.9 b), and Williamson (Fig. 3.9 c) counties. All three show plow damage and streaking. The one from Travis County has typical platform treatment (the platforms of the others were not photographed). Very deep patination, consistent with an inferred Clovis age, is revealed by the plow-damaged area near the distal end of the Williamson County specimen. All three are typical in the lack of negative bulbar scars on the core faces.

Two other surface finds, one (Fig. 3.10) from 41BX998 near San Antonio and another (Fig. 3.11) from the Gault site in Bell County (41BL323), retain evidence that might be indicative of how these cores were held or supported during blade removal. Both of these are conical cores with slightly elliptical platforms. Each of these cores has two opposing faces (Figs. 3.10 a, c and 3.11 a, c) and two opposing edges (Figs. 3.10 b, d and 3.11 b, d). The faces are moderately convex with multiple blade scars, and the edges have more prominent ridges formed by the convergence of the core faces. What is of interest on these cores is heavy crushing and bruising of these edge ridges (especially noticeable in Figs. 3.10 b and 3.11 b). It is clear that blade removals have been made from the two faces on each of these cores, and it is apparent that the dulled edges would have been in contact with whatever held or supported the core during knapping. Two possibilities present themselves. The edges may have been intentionally dulled by the knapper either to facilitate holding the core by hand (or between the feet) or improving its fit in an inanimate holding device. Alternatively, the dulling of the edges of these cores may have occurred inadvertently during indirect-percussion (punch) knapping if the core were in contact with some form of hard support or holding device. In either case, these cores seem to attest to concern with holding the core steady—a concern shared by prehistoric and contemporary blade knappers.

Of course, another alternative is that these cores were considered depleted or ruined (note the several hinge fractures on the core faces) and the edges retain damage sustained in some heavy form of usage that followed abandonment of the cores.

The place of blades in the technology of Clovis has received comparatively little notice by archeologists, who have emphasized bifacial lithics (particularly points) and artifacts of bone and antler. Even when considering sites or assemblages with blades, some writers have relegated Clovis blade technology to minor status and others have mistakenly assumed that true blades were blanks for the manufacture of points. Most Clovis blades are too curved, too narrow, and too asymmetrical in cross section to ever have served as the starting point for the manufacture of points. A few examples of the treatment of blade technology in the Clovis literature are instructional.

Judge's treatment of Paleoindian lithic technology twenty years ago reflects some common misconceptions of the time. First is the notion that there existed a sort of generalized Paleoindian lithic technology, and second is the spurious issue of blades as blanks for points. He also sees blade technology as relatively unimportant.

Flake blanks for the production of Paleoindian points were generally struck as very large flakes or blades. Although we hear much about the use of blades and the blade technique by various Paleoindian groups (especially Clovis), I feel this is somewhat of an over-reaction to the quest for specific Old World ties. I would suggest that most point blanks were originally flakes rather than blades, based on the existing evidence we now have from the Folsom data. . . . Of the various Paleoindian assemblages, I would suggest that only Eden points might have been derived from blade blanks. (Judge 1973: 88–89)

More-recent literature recognizes the distinctiveness of several Paleoindian lithic technologies, especially those of Clovis and Folsom. There is also growing recognition of blade production in Clovis technology, but it still receives less emphasis than does biface production.

Sanders (1990) illustrates and describes blades and blade cores from the Adams site in Kentucky but strongly emphasizes the bifacial technology of point production. He does, however, recognize the need for fuller investigation of Clovis blade technology in the eastern United States (Sanders 1990: 67).

Huckell (n.d.: 87–97) describes the blades, tools made on blades, heat-damaged blade core, and some blade-core debris at the Murray Springs site, his "blade and blade tool subsystem," and concludes with the observation,

The point to be stressed is that there is accumulating evidence that the Clovis industry includes the manufacture of true blades from prepared cores. While these are definitely not the central feature of the industry, neither are they flakes "...incidentally removed from bifaces during the normal course of reduction" (Callahan 1979:89). Additional data are needed to permit a more complete understanding of this subsystem, but there should be no doubt as to the existence of blade manufacture as part of the Clovis industry in the western United States.

The gradually increasing recognition of blades as part of Clovis assemblages is nowhere better illustrated than with a specimen from the Sheaman site in Wyoming (Frison 1982: Fig. 2.92 f). In the original site report, Frison describes the specimen as one of "five large biface reduction flakes" but nine years later cites that same specimen as evidence that "true blades are found occasionally in Clovis sites" (Frison 1982: 151; 1991: 321–322).

Stanford (1991) illustrates and discusses Clovis blade cores, blades, and tools made on blades in as much detail as bifacial stone artifacts and objects of bone and antler. It is telling that much of the evidence on blades that Stanford cites was, in 1991, either unpublished or minimally published.

With the current interest in the origin of Clovis, it is puzzling that its blade technology, which is so strikingly similar to those which preceded it in Europe, the Near East and Siberia, has not been more closely studied. The production of blades represents an elevated level of skill in the knapping of stone, but it is not so complex that multiple independent inventions are improbable. Users and makers of stone tools would readily appreciate the advantages of having relatively straight, sharp edges on a piece with the strength afforded by the triangular, prismatic, or trapezoidal section of a blade. Furthermore, the physical constraints of producing such forms would tend to impart common traits to historically unrelated blade-production technologies. Finding and convincingly demonstrating historical antecedents of Clovis blade technology, if they exist, will not be an easy task. A first order of business is fuller appreciation of this aspect of Clovis culture in the Americas.

Since Clovis blades were sometimes cached alone or with other kinds of artifacts, their study must address caching behavior along with matters of cultural history, lithic technology, and function.

PART TWO
The Keven Davis Site, a Clovis Blade Cache

I N SEPTEMBER 1988, KEVEN DAVIS, A SURVEYOR, discovered three fragmentary prismatic chert blades exposed by heavy earth-moving equipment near Cedar Creek in Navarro County, Texas. He brought the find to the attention of Bill Young of Corsicana, who is a regional steward in the Texas Archeological Stewardship Network, a volunteer organization affiliated with the Office of the State Archeologist at the Texas Historical Commission. Davis, Young, and Bobbie Jean Young returned to the site on several occasions and conducted further investigations. Their efforts recovered additional specimens and generated baseline information about the site, which is now beneath the waters of Richland-Chambers Lake.

The archeological community is indebted to Keven Davis, Bill Young, and the Texas Archeological Stewardship Network program for ensuring that the artifacts from, as well as information about, this important site were preserved and made available for this and future analyses.

Study of the artifacts from the site has been conducted under Texas Antiquities Permit No. 767. The collection was donated by Keven Davis to the Texas Historical Commission in Austin, where it is being housed. Two sets of casts made by Christopher Ferguson, of Santa Fe Replicast, Santa Fe, New Mexico, of all but one blade from the site also are curated at the Texas Archeological Research Laboratory and the Historical Commission.

The Keven Davis Cache (41NV659) at the time of its discovery was an isolated grouping of prismatic blades in an area no more than 2 meters across, and there were no associated archeological materials or other features to indicate the presence of a more complex site. Such are the circumstances indicating that these artifacts had been cached. The location is in southeastern Navarro County on a low interfluvial ridge between Cedar Creek and Little Cedar Creek, roughly 4 kilometers above (northwest of) the confluence of Little Cedar Creek and the much larger Chambers Creek (Fig. 4.1). This setting is about 0.6 kilometers north of Little Cedar Creek and 1.2 kilometers south of Cedar Creek. Cedar Creek valley is the wider and deeper of the two, and although the site is closer to Little Cedar Creek, it overlooks Cedar Creek and is only about 200 meters from a small, intermittent, unnamed tributary of that stream. Interpretation of the Keven Davis cache is dependent upon an understanding of the soil at the site and how it has been modified by farming and borrowing in this century.

The landscape of the area is low rolling hills supported by bedrock of the Wills Point formation, which belongs to the Paleocene-age Midway Group (Barnes 1988). These are soft, alkaline, silty and sandy marine clay rocks that weather to form a clay-rich soil, classified at the find spot as part of the Crockett series (Meade, Chervenda, and Greenwade, 1974: 2, 12–13, 65, and accompanying maps). Crockett soils are deep, Udertic Paleustalfs with a brown, fine sandy loam surface layer (typically 17 cm thick on level ground or 5 cm thick on slopes) overlying a deep layer of very firm clay, generally 105 to 170 cm thick. Crockett series soils have extreme shrink-swell characteristics with pronounced vertisolic structures. This material becomes extremely hard and deeply cracked when dry and very soft and sticky when wet. Samples of the clayey subsoil examined in the course of this study (see below) were found to contain rounded, sand-sized granules of manganese and iron along with fine to very fine sand grains of jasper and quartz.

Physically the soil at the site is notable for its extremes of hardness and vertical cracking when dry, plasticity and stickiness when wet. Chronologically, it is a very old soil that has certainly been in place throughout the human history of North America, having formed directly on weathered Tertiary bedrock.

Modern native vegetation on the site was typical of the Post Oak Savannah (Gould 1975; Kuchler 1964) until the land was cleared and brought into cultivation some seventy years ago; it remained in cultivation for an estimated forty years (Young 1988). Land clearing again in 1988 in anticipation of the filling of newly completed Richland-Chambers Lake included stripping of the loamy topsoil off of a large area at and around the site and hauling it to a planned residential subdivision nearby. It was this borrowing that exposed the blade cache and led to its discovery by Davis. The long history of cultivation of the locale followed by complete removal of the topsoil precluded any study of the intact soil profile where the blades were found. The blades were also heavily damaged by these activities.

At the time the cache was discovered and during subsequent investigations, the locality had lost most of its natural character (Fig. 4.2).

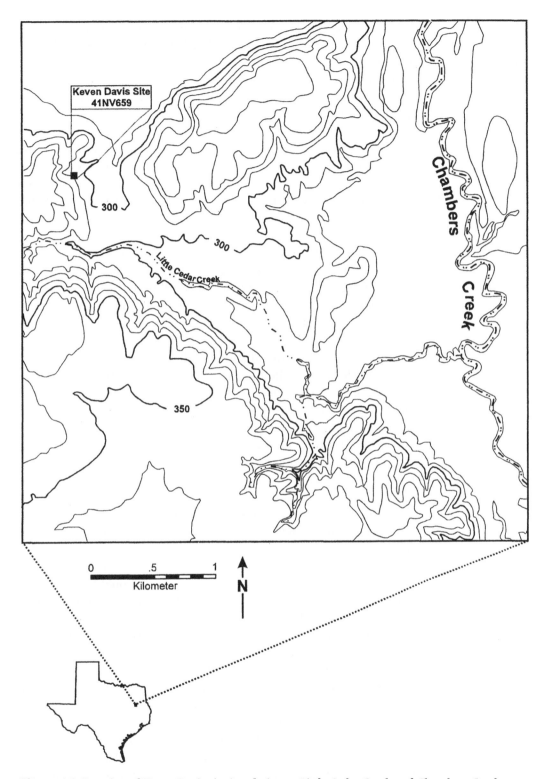

Figure 4.1. Location of Keven Davis site in relation to Little Cedar Creek and Chambers Creek.

The discovery occurred after the area had been prepared for impound-ment of Richland-Chambers Lake but before it was inundated. Key reference markers for this discussion are shown on the accompanying sketch map (Fig. 4.3) and consist of Lakeview Estates subdivision, a lake wall and boat ramp, abandoned county road, contoured shoreline, and borrow features.

The Keven Davis Cache at the time of its discovery had spatial dimensions of two kinds. First was the grouping of artifacts ("the cache") in an area less than 2 meters across lying within the impoundment area. Second was this cache spot plus the area where pieces inferred to be from the cache were dispersed along with fill dirt in the Lakeview Estates subdivision. This greater dimension defines a locality that straddles the shoreline. Preparation of the shoreline along the waterfront in Lakeview Estates included the cutting of a steeply sloping bank some 4 meters high and the installation of a lake wall and boat ramp (Figs. 4.2 and 4.3). Lakeward from this cutbank was a strip about 55 meters wide where borrowing extended more than 25 cm into the subsoil, in places as much as ca. 3 meters. This fill was spread on the adjacent lots. Farther from the cutbank was a wider strip where generally around 25 cm of fill was removed and spread on the nearby lots; the cache was exposed in this strip. At the time Davis and Young conducted their work at the site, small ridges of soil ori-ented northwest-southeast in this second strip (Fig. 4.2 b) indicated that power equipment (scraper pans?) had operated more-or-less par-allel to the shoreline (that is, northwest-southeast). This second strip was approximately 65 meters wide. Borrowing had not extended beyond this strip.

At the time I visited the site in February 1989, the surface of both stripped areas was clayey subsoil, saturated from recent rains, and the more-remote area where borrowing had not occurred was clearly dis-turbed from years of cultivation. In this saturated condition, the sub-soil was extremely plastic and sticky, and although mapped as loamy, the plowzone, too, was very plastic and sticky. It would be impossible to conduct borrowing operations in this area if the ground were satu-rated. On the other hand, if it were completely dry, scraper pans could not cut into unbroken ground. This leaves two possible scenarios for the borrowing activities that exposed the cache if scraper pans were used. Either the ground was slightly moist and could be effectively borrowed through the use of scraper pans alone or, if the ground were too hard, chisel plowing with a bulldozer had to precede scraping. In another alternative, if the soil were dry, borrowing may have been accomplished with a crawler-loader and dump trucks, in which case, chisel plows behind the loader were probably used. An important dif-ference between picking up loosened earth with a scraper pan and

Figure 4.2. The Keven Davis site as it appeared in February 1989: (a) view from the south, showing stripped area (Figures are standing near the excavations at the original find spot; note eroded toe of the sloping bank at left foreground); and (b) closer view from the south of excavated area (note wet, plastic nature of subsoil and ridges of dirt left by the passing of earth-moving equipment).

with a crawler-loader would be that the former travels on rubber tires and the latter on metal treads. Also, if moisture conditions were optimal for use of scraper pans, chisel plowing would not have been necessary. Field evidence, in the form of the long, straight ridges of spoil dirt that are typical of those left by these machines (Fig. 4.2 b) and the absence of bulldozer tracks and chisel-plow scars, favors the scraper-pan method without use of other equipment. These issues pertain to the circumstances in which the cached blades were broken.

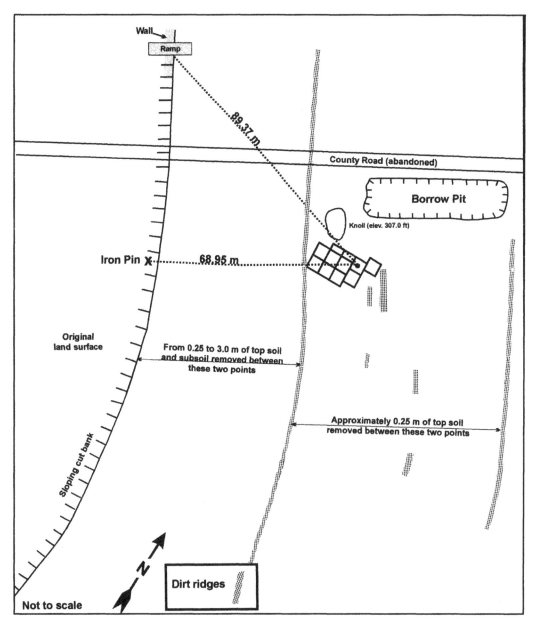

Figure 4.3. Sketch map of the Keven Davis site (41NV659).

Field and Laboratory Investigations   FIVE

I NITIAL DISCOVERY OF THE SITE OCCURRED ACCIDEN-
tally on 18 September 1988 when Davis was surveying in the
Lakeview Estates subdivision. Three prismatic blades exposed on
the bare surface of the borrow area were noticed and collected. After
Young learned of the find and visited the site, surface examination,
mapping, and limited excavations were conducted, the latter extending into October 1988. Young and, especially, Davis repeatedly
searched the area where borrowed dirt was spread. Davis found two
additional pieces in July, 1991. Because no other kinds of artifacts have
been found and because the blades are all so similar, it is inferred that
they were originally cached together. One fragmentary blade, more
deeply patinated than the others and found in the area of dispersed
fill, may or may not be an exception to this inference.

Young and Davis gridded a small, irregular excavation block of thir-   **EXCAVATIONS**
teen 1-meter squares centered over the spot where the blades were
found originally. This grid was oriented with magnetic north, some 21
degrees from the alignment of the road (see Figs. 4.3 and 5.1). They
shovel-scraped and screened loose dirt from the area where the first
three blades were found and were able to document the location of
five chert pieces at or near the surface. This was followed by excavations in which nine complete and two half squares of the thirteen were
excavated and another sixteen pieces recovered. Depth of excavation
ranged from 30 to 50 cm below surface. Fill removed from these
squares was passed through screens with 1/4-inch-mesh wire. Two
samples of soil, each ca. 2.5 liters in volume, were collected in the
vicinity of the greatest concentration of blade fragments, and three
smaller samples were taken from other positions in the excavation.
The two larger soil samples and portions of two smaller ones were
water-screened by me through fine mesh (0.71 mm) wire in the

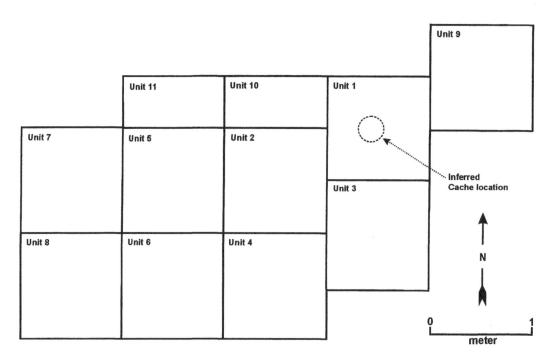

Figure 5.1. Plan map of completed excavation units.

laboratory. Blade segments found in place in the excavation were documented for vertical and horizontal location as well as for position.

When these distributional data are examined closely, some insight is gained into the forces of disturbance that have acted upon the cache. As noted above, the borrowing operation that exposed the cache was accomplished by equipment (probably rubber-tired scraper pans) moving on a northwest-southeast alignment. When the positions of conjoinable pieces are connected with straight lines (Fig. 5.2), it is clear that a disturbance trajectory at very nearly right angles to this borrowing alignment had a major effect on this cache—that is, from northeast to southwest. Overall, the scatter seems to fan out in a southwesterly direction from the small area of concentrated fragments over which grid square 1 was centered (Fig. 5.2). That concentration is inferred to have been the original cache position, and its dispersal pattern is almost identical to that documented by Mallouf (1982: 81 and Fig. 2) for the Brookeen blade cache.

The damage on the Keven Davis pieces also matches closely that which Mallouf (1982) documented on the Brookeen pieces. It is important to note that there are two degrees of freshness to the surfaces on the blades and blade fragments from the Keven Davis cache (not including the patinated segment). Overall, the material is not

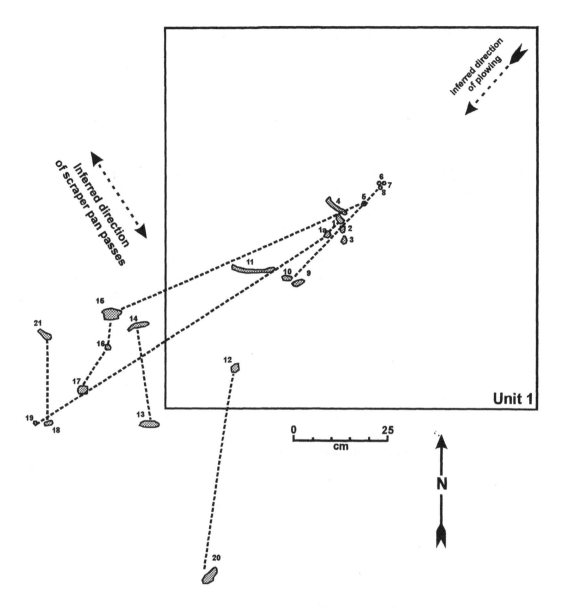

Figure 5.2. Plan map of piece-plotted artifacts, by piece number, in and near excavation unit 1. Dashed lines denote refitted pieces and define a primary vector of disturbance from northeast to southwest at roughly a right angle to the northwest-southeast orientation indicated for passage of equipment that exposed the artifacts in 1988.

patinated and the few nicks that occurred during archeological recovery are only slightly fresher in appearance. The remainder of breaks and edge damage have the same virtually fresh appearance as do the blade surfaces. These observations point to the conclusion that plowing at some time when this area was in cultivation (probably between 1920 and 1950) was responsible for the breakage and that only when the borrowing occurred in 1988 were the pieces exposed.

ANALYTICAL METHODS

Shortly after the field recovery efforts, Young and Davis washed all of the pieces recovered during their excavations and systematically attempted to refit broken pieces and glue them together. The chert specimens were handled extensively in the field and during subsequent washing and study, thus precluding the extraction and identification of trace organic residues. They used a retail water-soluble white glue ("Elmer's" brand) to affix conjoinable pieces and were able to make twelve major refits. Young executed good pen-and-ink drawings of each blade or blade segment after these repairs were made. His drawings are part of the site records for this site (on file, Texas Archeological Research Laboratory, The University of Texas).

During my later study of this collection, I water-screened all but a small part of the matrix samples from the Young and Davis excavations. This was done through two nested standard sieve screens of 1.41 mm and 0.71 mm mesh.

The first sample, approximately 0.25 liter in volume, bore the field notation, "unusual organic, 41 cm deep, Unit 1, 5 S and 21 E of Center Point" (meaning 5 cm south, and 21 cm east, of the center point of unit 1). The second sample, approximately 5 liters in volume, was labeled, "Soil Sample where 1st 4 blades were found."

It was necessary to soak the matrix in a solution of tri-sodium phosphate and water to deflocculate the clay aggregates before screening. The detritus collected on the top of these two screens was systematically examined under low-power magnification (10 and 30 power).

There was no recovery of chert from the first sample. Screening of the second sample resulted in the recovery of five fresh chips of Chert 1, all clearly produced in the breakage of pieces in this cache. Three of these were recovered on the 1.41 mm screen, lacked cortex, and were in the 3 to 4 mm size range. Two were found on the 0.71 mm screen; one is a slender sliver 4 mm long, and the other is a cortical chip 2 mm long.

Two 55 ml subsamples, "a" and "b," of the first, "unusually organic," soil sample were completely dissolved in water and centrifuged. These swelled from a dry volume of 55 ml to 83.8 ml for "a" and 92.5 ml for "b" when completely dissolved in water and, when centrifuged, produced the following measures:

Subsample "a"
sand—3.5 ml (4%)
silt—11.5 ml (14%)
clay—68.8 ml (82%)

Subsample "b"
sand—2.0 ml (2%)
silt—15.0 ml (16%)
clay—75.5 ml (82%)

The "sand" in these samples consists of grains of quartz and jasper along with small granules and concretions of manganese and iron. These data underscore the extremely high clay content of the matrix in which the cache occurred.

In order to facilitate microscopic examination of the blades for use wear and to evaluate the breakage, it was necessary to loosen the glue joints made by Young and Davis. This proved difficult, as the "water soluble" glue was extremely tenacious, even in very warm water. Kay (see Chapter 7) was able to loosen the glue by placing each restored blade or blade segment in a plastic bag with mild detergent, immersing the bag in warm water in an ultrasonic cleaner, and operating the cleaner until the glue joints loosened (up to one half hour, in some cases).

After the high-power microscopic examination (see Chapter 7) and macroscopic inspection of all of the unglued fragments, the specimens were reglued. This second gluing was done with B72 (a product of H. Marcel Guest, Ltd.), which is a commercial, acetone-soluble adhesive of the kind recommended for archeological materials (Koob 1986); this adhesive is easily reversed using acetone. Unfortunately, the tube used in this effort was not entirely fresh, and the results were not as satisfactory as they might have been. However, because of the ease with which this glue can be removed and replaced, the pieces can be refitted another time if necessary.

Finally, a systematic attempt was made to refit interior and exterior blade scars to identify blades removed consecutively. Every possible combination was tried repeatedly, but only one set of consecutive blades in Chert 1 and none in Chert 2 was identified.

In spite of efforts to standardize terminology in lithic analysis, there are still numerous inconsistencies and ambiguities in the terms used, so a glossary defining the technical vocabulary used here appears at the end of this book.

Thirteen measurements and indices were recorded on the Keven Davis blades (Fig. 5.3 and 5.4). These are defined as follows:

maximum length: *straight-line length in mm from the platform to the most distal point on the blade, taken with sliding calipers (Fig. 5.3).*
maximum width: *the widest point in mm on the blade between the lateral margins, taken with sliding calipers (Fig. 5.3).*

maximum thickness: *taken at the point of greatest thickness, as measured in mm from the interior to the exterior surfaces with sliding calipers (Fig. 5.3).*

platform angle: *the angle between the proximal interior blade surface and the platform (Fig. 5.3), measured visually by placing the blade on a template (Fig. 5.4); it is not possible to precisely determine the plane of very small platforms, which are characteristic of many of these blades, so this measurement is not to be considered as more accurate than ± 5 degrees (that is, the true angle falls within a 10-degree range).*

weight: *measured directly on electronic scale in grams.*

platform width: *the maximum dimension in mm of the striking platform between the lateral edges, taken with sliding calipers (Fig. 5.3).*

platform depth: *the maximum dimension perpendicular to the platform width, from the interior to the exterior surfaces, taken in mm with sliding calipers (Fig. 5.3); this represents the "bite" taken by the punch or the percussor on the platform of the core.*

index of curvature: *the ratio of two linear measurements taken on the interior surface of the blade; these measurements are (a) the straight-line distance between the distal and proximal points of contact of the interior blade surface and a flat plane, and (b) the maximum perpendicular distance between that plane and the interior surface of the blade (Fig. 5.3); the greater the value of the index, the more curved the blade; since this is a ratio, it can be calculated on an incomplete blade segment. This is not considered a highly precise measure but, rather, a generalized expression of curvature for descriptive and comparative purposes.*

width-to-length ratio: *the arithmetic expression of the maximum length in relation to maximum width, with width given an arbitrary value of one.*

length + width + thickness: *the sum of the measurements maximum length, plus maximum width, plus maximum thickness; this is calculated for use in further calculating the three ratios next described; by itself, this has little value other than as a very generalized expression of overall size.*

length divided by length + width + thickness: *the ratio of length to the sum of the primary dimensions, used in graphic presentation of shape.*

width divided by length + width + thickness: *the ratio of width to the sum of the primary dimensions, used in graphic presentation of shape.*

thickness divided by length + width + thickness: *the ratio of thickness to the sum of the primary dimensions, used in graphic presentation of shape.*

Figure 5.3. Measurements, attributes, and landmarks considered in this study of Clovis blades.

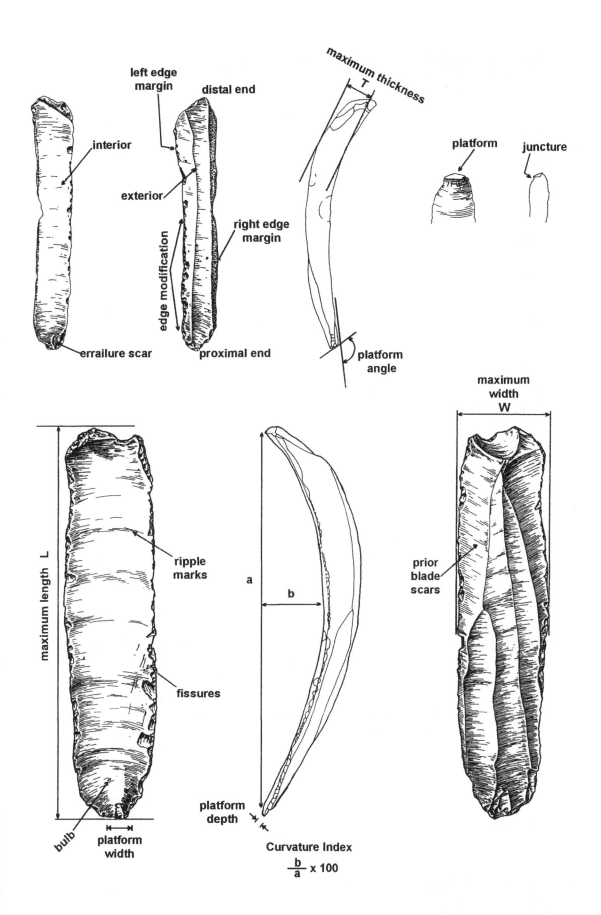

Figure 5.4. Method of estimating the platform angle. Note that the platform surface on most Clovis blades is very small and is sometimes convex as well, making this estimate imprecise.

Macroscopic, nonmetric variables considered in this study are primarily technological features of the blades (Fig. 5.3). The exterior and platform surfaces of the blade retain characteristics of the core prior to blade detachment, whereas interior surface attributes and size were imparted to the blade at the time of its removal. The observed attributes and their states are here described individually.

COMPLETENESS   The first observation considered the overall blade for completeness. In most cases, it is obvious whether the specimen is a segment of a blade or an essentially complete blade. However, because of the complex and highly variable way in which blades can terminate at the distal end of the core, it is not always possible to distinguish minor, postdetachment breakage from original termination. Thus blades are here described as complete unless the platform is missing or the distal termination is an obvious break. A case in point is described below in which blade 1 terminates with a diving, rounded facet on its exterior, apparently reflecting the configuration of the distal end of the core.

Late in the analysis of the collection, a small piece long thought to represent another blade (assigned number 12) was found to fit onto blade 1, changing this interpretation but not significantly changing the configuration of the blade. In the case of broken blades, the configuration of the fracture plane was compared to those described by Mallouf (1982) in an effort to separate machine damage from other kinds of breaks.

The exterior surface of each blade was observed for presence, extent, and nature of cortex. The noncortical facets were examined for direction of detachment, shape (whether blade- or flakelike), and the presence or absence of negative bulbs. Each ridge between facets was observed for small scars, battering, rounding, or other modifications. Blade edges were examined for modification, and inferences were made as to the origin of any modifications present. On most blade edges, some flaking and nicking is present. The primary consideration is whether this was caused by machinery in recent times or was the result of intentional retouching or unintentional use-damage prehistorically. Comparison with the machine-damaged pieces described by Mallouf (1982) and microscopic observations for telltale metal streaks were the primary means for making these inferences.

EXTERIOR

Blade interiors were examined for size and shape of the bulb of percussion, presence or absence of ripple marks, and presence or absence of erraillure scars. Where marginal edge modification intrudes onto the interior surface, it was observed and is described in the same manner as that which extends onto the exterior surface. Glenn Goode has found that low-amplitude ripples can be produced from either direct, soft-hammer percussion or from pressure techniques (see Chapter 2).

INTERIOR

Each blade platform was examined for its overall geometry, degree of faceting (single, double, multiple), and any evidence of dulling.

PLATFORM

The proximal edge of the interior surface of each blade was examined for the presence or absence of lipping.

PLATFORM/INTERIOR JUNCTURE

The fracture planes of incomplete blades were examined for evidence of the cause of breakage. The kind of fracture present was compared to well-defined breaks of known cause, such as "perverse" (Crabtree 1982:46) or bend fractures. Bend fractures are the most common kind of break observed on the Keven Davis blades. It is of interest to know the direction in which force acted to bring about each of these bend breaks. Since breaks of this kind have a distinctive lip along one margin of the fracture plane, they can be oriented. In the

FRACTURES

literature consulted for the interpretation of bend breaks, diametrically opposite diagnoses were offered: Mallouf (1981, 1982) discusses and illustrates bend breaks and identifies the lip as forming on the outside of the bend, whereas Whittaker (1994) shows the lip as forming on the inside of the bend. To resolve this, I subjected thirty-four experimental blades and flakes to bending force of the kind that might be expected under the weight of heavy equipment passing over a piece lying horizontally in the ground, or from a metal plow blade moving through the ground and hitting a piece near midpoint. These situations would resemble a three-point application of force with two separated vectors in one direction and an intermediate vector in the opposite direction (Fig. 5.5). All thirty-four of these experimental breaks produced the lip on the inside of the bend, and it is inferred that the bend breaks on the Keven Davis blades are analogous.

LOW-POWER MICROSCOPY

A binocular 10X to 70X zoom microscope was used to enhance some of the observations just described. This was particularly beneficial in ascertaining whether grinding had been used to prepare a platform before blade removal and in examining ridges on the exterior of a blade for bruising or microchipping.

CONTEXT

In the descriptions of the individual blades, the location of the piece or pieces in the site is noted.

Combining these observations and seriating blades according to the order in which they were removed permits the recognition of six stages of removal. Not all of these stages occur in the Keven Davis blades, but they are well represented in some of the comparative assemblages—especially those from the Pavo Real Site (see Chapter 3). These stages are as follows:

Stage 1. *Primary blades with natural exterior surfaces; natural exteriors are cortex or partially cortex; lateral edges are irregular; both bulbs and platforms are often large, suggesting hard-hammer, direct percussion; longitudinal sections are straight to slightly curved.*

Stage 2. *Secondary cortex blades with prepared crests and possibly one or two scars of prior blade removals, usually multiple flake scars of varying orientations; edges irregular, usually with flake scars; slight longitudinal curvature in most specimens, but moderate curvature in a few; large bulbs, possibly produced by hard-hammer percussion.*

Stage 3. *More-regular blades with minor cortex or crest remnants; usually one or two, sometimes more, prior blade scars on exterior; core-preparation flake scars present on exterior; edges irregular; flat bulbs and large bulbs both occur; variable, but generally relatively little, longitudinal curvature.*

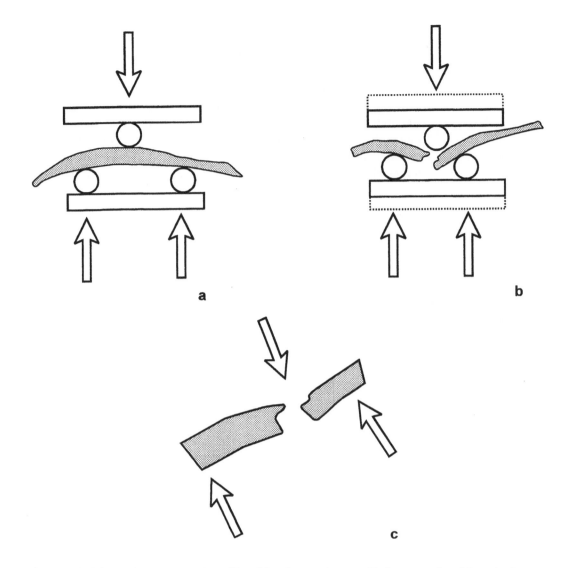

Figure 5.5. Schematic representation of bend-break experiments. Blades were placed in a vice in contact with (a) three soft wooden dowels, and the three points of contact were marked on the blade with an indelible marker. The vice was closed until (b) the blade snapped. The marked points of contact were then examined in relation to (c) the bend break, and consistently it was found that the lip was toward the inside of the bend as shown.

Stage 4. *Moderately regular blades, little or no cortex, multiple prior blade scars and some core-preparation flake scars, flat bulbs, moderate curvature.*
Stage 5. *Regular blades, no cortex, multiple prior blade scars, flat bulbs, strong curvature.*
Stage 6. *Very regular blades, no cortex, multiple prior blade scars, flat bulbs, strong curvature; some scars of prior blades removed from distal end of core (some relatively narrow blades are seen in this stage).*

These stages are not absolutely sequential, and the distinctions between stages are somewhat subjective. They do provide a generalized indication of where a given blade fits in the sequence of removals in an idealized blade-core reduction. One needs to keep in mind that the face of a Clovis blade core often wraps around the nodule completely or partially, and one portion of the face may be in a more advanced stage of reduction than another. This means that, for example, a stage 6 blade could be detached from one part of the core, followed by removal of a stage 2 blade from another part of the core.

S URFACE COLLECTING, EXCAVATION, AND LABORA-
tory screening combined yielded twenty-seven blade segments
and nearly complete blades (Fig. 6.1) along with a number of
small chips, twelve of which conform in color and texture to the cache
pieces(Fig. 6.2). This includes the one patinated blade segment found
on the surface that may or may not have been a part of the cache.
When twenty-one fragments among these twenty-seven pieces are
refitted (Fig. 6.3), ten essentially complete blades and four blade segments can be identified; since none of these four fragments appear to
be from the same blade, a total of fourteen blades is represented from
the site. (Young and Collins [1989] reported fourteen blades in a preliminary note on this cache, prior to the 1991 discovery of an additional blade; two fragments thought at the time of that publication to
be pieces of separate blades were subsequently found to fit together,
reducing the total by one.) The fourteen blades are described individually below. Metric and other data are summarized in Table 6.1.

The blades in this cache are of two kinds of high-quality chert,
herein referred to as Chert 1 and Chert 2. Ten certainly, and probably
eleven, are of Chert 1, which is a slightly brownish, light gray, banded
Edwards chert with a creamy white, soft cortex. This material is
homogeneous and lacks any flaws such as crystal masses, seams, or
internal fractures. It is opaque and moderately lustrous. It fluoresces,
but not brightly, to yellowish olive green under both short- and long-
wave ultraviolet light. In all of these characteristics, it resembles a
variety of Edwards chert (Banks 1990) found in west-central
Williamson County and often referred to as "Georgetown chert." The
outcrop area of this distinctive material is geographically restricted,
and the identification of this as the source of the material used in
the manufacture of most of the Keven Davis blades can be made with

**RAW-MATERIAL
DESCRIPTIONS**

Figure 6.1. The twenty-seven larger pieces making up the Keven Davis cache find. Designations of these pieces are shown schematically, below the plate.

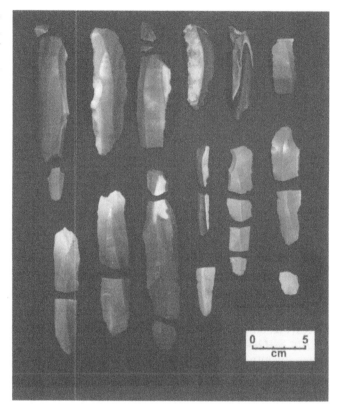

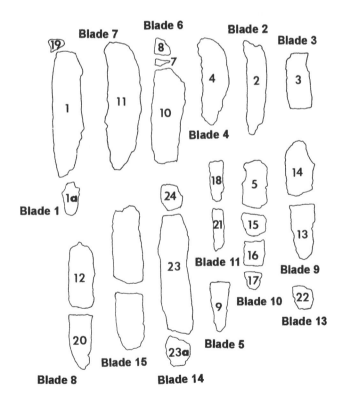

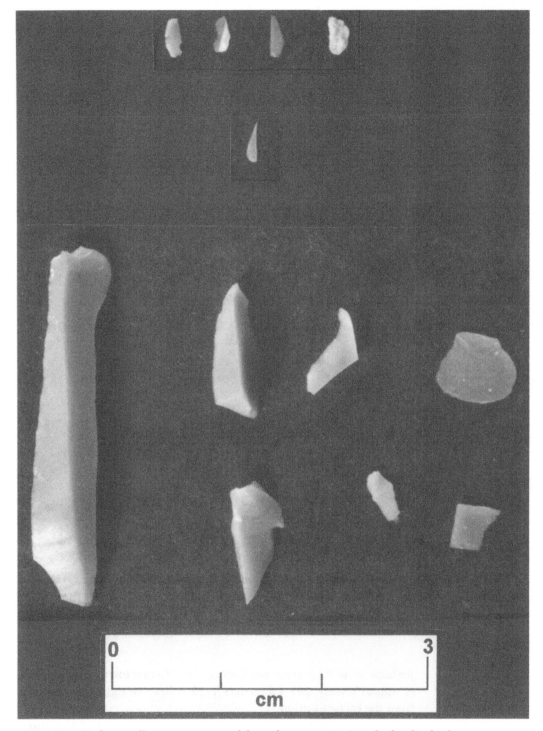

Figure 6.2. Twelve smaller pieces recovered from the Keven Davis cache locality by fine-screening.

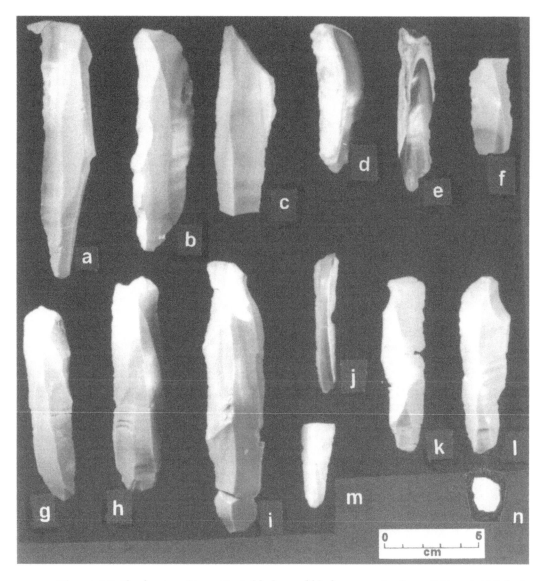

Figure 6.3. The fourteen Keven Davis blades and blade segments after refitting: (a) blade 1, (b) blade 7, (c) blade 6, (d) blade 4, (e) blade 2, (f) blade 3, (g) blade 8, (h) blade 15, (i) blade 14, (j) blade 11, (k) blade 10, (l) blade 9, (m) blade 5, and (n) blade 13.

perhaps 90 to 95 percent confidence. The "Georgetown" outcrop area is approximately 205 kilometers (125 miles) straight-line distance from the cache location.

Chert 2 is unidentified as to source and is the material in three blades from the cache. The cortex of Chert 2 is soft and superficially appears to be light reddish brown in color but, on closer inspection,

BLADES FROM CLOVIS CONTEXTS

	ART. #	MAX L	MAX W	MAX T	PLAT.>	WT g	PLAT W	PLAT D	INDEX CURV	RATIO W:L	L+W+T	L/L+W+T	W/L+W+T	T/L+W+T	APPROX. % CORTEX	STAGE
KEVEN DAVIS	1	139	28	16	110	50.0	8.3	2.7	15.2	4.97	182	0.76	0.15	0.09	0	6
	2	88	20	10	110	17.2	6.9	4.6	14.1	4.34	119	0.74	0.17	0.09	30	2
	3	*53	23	9	N/A	11.6	N/A	N/A	18.0	N/A	N/A	N/A	N/A	N/A	N/A	N/A
	4	80	26	12	130	26.1	9.4	4.2	10.0	3.15	117	0.68	0.22	0.10	50	2
	5	19	7	7	120	5.1	3.4	1.6	2.5	N/A	N/A	N/A	N/A	N/A	N/A	N/A
	6	102	32	9	N/A	26.6	N/A	N/A	16.3	N/A	N/A	N/A	N/A	N/A	12	4
	7	119	32	11	130	47.1	7.3	2.1	13.4	3.68	162	0.73	0.20	0.07	8	4
	8	105	25	13	110	26.9	5	2.2	16.5	4.13	142	0.73	0.18	0.09	0	6
	9	95	28	9	100	22.7	5.3	2.1	13.6	3.38	132	0.72	0.21	0.07	0	6
	10	94	24	9	120	18.7	5.4	2.1	14.8	3.91	126	0.74	0.19	0.07	30	2
	11	75	13	7	100	7.8	4.2	2	12.0	5.57	95	0.78	0.14	0.08	N/A	N/A
	13	*23	19	6	110	2.2	10.2	1.9	0.0	N/A	N/A	N/A	N/A	N/A	N/A	N/A
	14	*145	33	14	N/A	62.2	N/A	N/A	13.7	N/A	N/A	N/A	N/A	N/A	N/A	N/A
	15	118	29	13	N/A	40.7	N/A	N/A	16.2	4.07	160	0.74	0.18	0.08	0	6
MEANS		**101**	**25**	**10**						**4.04**	**137**	**0.74**	**0.18**	**0.07**		
BLACKWATER DRAW "A"	1	138	34	14	135				11.3	4.06	186	0.74	0.18	0.08	5	6
	2	156	33	12	136				12.2	4.73	201	0.78	0.16	0.06	0	6
	3	140	30	19	111				13.8	4.67	189	0.74	0.16	0.10	30	5
	4	140	31	15	142				14.8	4.52	186	0.75	0.17	0.08	0	4
	5	99	30	10	120				9.0	3.30	139	0.71	0.22	0.07	0	6
	6	103	29	12	130				13.3	3.55	144	0.72	0.20	0.08	20	5
	7	N/A	35	17	N/A				16.6						0	4
	8	N/A	33	11	N/A				12.5							
	9	N/A	29	19	N/A				12.5						25	2
	10	N/A	23	12	128											6
	11	N/A	33	11	N/A											1
	12	N/A	29	14	N/A											2
	13	N/A	34	14	N/A											1
	14	N/A	27	14	N/A											5
	15	N/A	30	11	N/A											3
	16	N/A	35	20	N/A											6
	17	N/A	34	12	N/A											5
MEANS		**129**	**31**	**14**	**129**				**12.9**	**4.16**	**174**	**0.74**	**0.18**	**0.08**		

TABLE 6.1. Metric data on blades from Keven Davis and selected sites.

ART. #	MAX L	MAX W	MAX T	PLAT.>	WT g	PLAT W	PLAT D	INDEX CURV	RATIO W:L	L+W+T	L/L+W+T	W/L+W+T	T/L+W+T	APPROX. % CORTEX	STAGE
BLACKWATER DRAW "B"															
	215	56	15			17	4		3.84	286	0.75	0.20	0.05		
	120	30	6			4	2		4.00	156	0.77	0.19	0.04		
	135	44	14			5	3		3.07	193	0.70	0.23	0.07		
	130	40	15			N/A	N/A		3.25	185	0.70	0.22	0.08		
	137	34	12			N/A	N/A		4.03	183	0.75	0.19	0.07		
MEANS	**147**	**41**	**12**			**9**	**3**		**3.61**	**201**	**0.73**	**0.20**	**0.06**		
RICHEY ROBERTS*															
	80	24	9	N/A				13.8	3.33	113	0.71	0.21	0.08		6
	68	24	6	N/A				13.4	2.83	98	0.69	0.24	0.06		5
	124	47	15	N/A				14.8	2.64	186	0.67	0.25	0.08		4
	85	32	8	N/A				14.6	2.66	125	0.68	0.26	0.06		5
	120	32	18	N/A				15.3	3.75	170	0.71	0.19	0.11		6
MEANS	**95**	**32**	**11**					**14.4**	**2.97**	**138**	**0.69**	**0.23**	**0.08**		
PAVO REAL															
	120	30	23	110		4	3	7.6	4.00	173	0.69	0.17	0.13	0	4
	127	25	11	110		11	4	2.5	5.08	163	0.78	0.15	0.07	0	6
	145	37	21	100		7	3	3.4	3.92	203	0.71	0.18	0.10	50	1
	153	37	21	110		21	12	1.3	4.14	211	0.73	0.18	0.10	50	1
	86	29	10	100		9	3	12.7	2.97	125	0.69	0.23	0.08	0	5
	77	28	12	100		18	4	10.0	2.75	117	0.66	0.24	0.10	0	5
	130	72	25	120		46	20	7.6	1.81	227	0.57	0.32	0.11	0	1
	103	29	18	100		19	10	8.0	3.55	150	0.69	0.19	0.12	45	5
	56	18	9	120		6	2	1.8	3.11	83	0.67	0.22	0.11	0	1
	102	31	18	100		11	7	10.0	3.29	151	0.68	0.21	0.12	90	1
	86	22	15	110		15	7	3.5	3.91	123	0.70	0.18	0.12	0	3
	N/A	44	15	100		10	5	2.5						0	3
	N/A	37	14	90		18	2	3.6						25	5
	N/A	44	21	110		11	3	5.0						35	4
	N/A	39	24	110		11	4	7.3						50	2
	N/A	31	14	110		6	3	3.3						50	1
	N/A	37	16	N/A		N/A	N/A	5.4						0	4
	N/A	21	6	N/A		14	2	4.2						5	4
	N/A	28	12	125		11	6	10.0						0	5
	N/A	18	7	N/A		N/A	N/A	5.0						0	6
	N/A	18	9	N/A		N/A	N/A	1.2						55	1
	N/A	22	8	N/A		N/A	N/A	5.7						0	6
	N/A	16	6	N/A		N/A	N/A	7.1						0	6

	N/A	14	5	N/A		N/A	N/A	7.4						0	6
	N/A	15	5	N/A		N/A	N/A	6.0						0	6
	N/A	33	13	110		22	10	0.0						0	5
	N/A	46	11	N/A		N/A	N/A	4.4						30	4
	N/A	42	10	100		14	6	10.0						20	4
	93	32	17	130		11	4	15.6	2.91	142	0.65	0.23	0.12		5
	N/A	22	7	110		18	4	9.5						30	2
	N/A	23	7	N/A		N/A	N/A	5.0						0	6
	78	31	12	100		13	7	0.0	2.52	121	0.64	0.26	0.10	0	5
	102	25	17	100		13	5		4.08	144	0.71	0.17	0.12	75	
MEANS	**104**	**30**	**13**	**108**		**14**	**5.6**	**5.8**	**3.44**	**148**	**0.71**	**0.20**	**0.09**		
MURRAY SPRINGS	87	29	6					16.5	3.00	122	0.71	0.24	0.05		
HORN 2		19	7			10	3	7.0							
FENN*	70	24	7						2.92	101	0.69	0.24	0.07		
ADAMS MEANS**	**82**	**33**	**12**		34				**2.48**	**127**	**0.65**	**0.26**	**0.09**		
PROBABLE CLOVIS CONTEXTS															
DOMEBO	86	32	10	N/A				0.0	2.69	128	0.67	0.25	0.08	0	5
GAULT 1	106	33	11	120		14.5	5.7	9.8	3.21	150	0.71	0.22	0.07	0	6
GAULT 2	N/A	17	15	N/A				7.4						0	6
GAULT 3	N/A	21	18	110		2.2	1.2	22.5						0	6
GAULT 4	N/A	32	10	N/A				12.0						0	5
GAULT 5	N/A	19	5	N/A				N/A						N/A	6
GAULT 6	N/A	42	17	N/A				0.0						0	3
MEANS	**106**	**27**	**13**	**115**		**8.4**	**3.5**	**10.3**	**3.93**	**146**	**0.73**	**0.18**	**0.09**		
McFADDIN 62	105	32	12						3.28	149	0.70	0.21	0.08		
McFADDIN 42	100	30	10						3.33	140	0.71	0.21	0.07		
McFADDIN 113	63	23	8						2.74	94	0.67	0.24	0.09		
MEANS	**89**	**28**	**10**						**3.18**	**127**	**0.70**	**0.22**	**0.08**		
SPRING LAKE	79	20	8	130		6	2	19.5	3.95	107	0.74	0.19	0.07	0	5
RN 107	100	30	14	120		9.5	5.5	8.5	3.33	144	0.69	0.21	0.10	20	2

	ART. #	MAX L	MAX W	MAX T	PLAT.>	WT g	PLAT W	PLAT D	INDEX CURV	RATIO W:L	L+W+T	L/L+W+T	W/L+W+T	T/L+W+T	APPROX. % CORTEX	STAGE
CROCKETT GARDENS*		93	24	14	N/A	N/A	N/A	N/A	6.5	3.88	131	0.71	0.18	0.11	0	6
		52	21	8					0.0	2.48	81	0.64	0.26	0.10	0	5
		94	28	10					5.3	3.36	132	0.71	0.21	0.08	0	4
		N/A	26	16					13.3		N/A	N/A	N/A	N/A	0	6
		N/A	25	9					7.8		N/A	N/A	N/A	N/A	0	5
MEANS		**80**	**25**	**11**					**6.6**	**3.20**	**116**	**0.69**	**0.22**	**0.09**		
INDEFINITE CONTEXTS																
CEDAR CREEK	a	102	47	13	128				7.5	2.17	162	0.63	0.29	0.08	0	5
	b	90	36	12	110				14.3	2.50	138	0.65	0.26	0.09	0	4
	c	106	35	14	N/A				2.5	3.03	155	0.68	0.23	0.09	10	4
	d	82	22	13	N/A				16.1	3.73	117	0.70	0.19	0.11	0	4
	e	103	33	11	115				15.0	3.12	147	0.70	0.22	0.07	0	4
	f	92	29	14	125				2.6	3.17	135	0.68	0.21	0.10	0	4
	g	110	36	11	105				5.0	3.06	157	0.70	0.23	0.07	10	4
MEANS		**98**	**34**	**13**	**117**				**9.0**	**2.88**	**145**	**0.68**	**0.23**	**0.09**		
ANADARKO	1	95	25	8	110				12.5	3.80	128	0.74	0.20	0.06	10	5
	2	95	29	8	120				17.5	3.28	132	0.72	0.22	0.06	0	6
	3	97	32	16	95				12.5	3.03	145	0.67	0.22	0.11	0	5
	4	132	35	10	105				9.0	3.77	177	0.75	0.20	0.06	5	5
	5	120	27	5	110				9.0	4.44	152	0.79	0.18	0.03	5	5
	6	122	25	7	95				12.0	4.88	154	0.79	0.16	0.05	0	6
	7	102	25	10	N/A				12.2	4.08	137	0.74	0.18	0.07	0	5
	8	120	37	11	115				12.2	3.24	168	0.71	0.22	0.07	0	5
	9	110	25	10	N/A				6.6	4.40	145	0.76	0.17	0.07	0	6
	10	106	31	7	N/A				10.0	3.42	144	0.74	0.22	0.05	0	5
	11	110	36	12	N/A				14.3	3.06	158	0.70	0.23	0.08	0	6
	12	109	36	8	112				7.1	3.03	153	0.71	0.24	0.05	0	5
	13	100	42	9	125				11.6	2.38	151	0.66	0.28	0.06	0	5
	14	108	37	9	120				2.2	2.92	154	0.70	0.24	0.06	10	3
	15	140	48	11	118				5.0	2.92	199	0.70	0.24	0.06	40	3
	16	106	46	8	N/A				11.4	2.30	160	0.66	0.29	0.05	20	3
	17	95	37	7	115				7.5	2.57	139	0.68	0.27	0.05	20	3
	18	114	48	13	120				4.4	2.38	175	0.65	0.27	0.07	0	4
	19	135	50	16	118				9.0	2.70	201	0.67	0.25	0.08	10	4
	20	105	48	7	100				2.5	2.19	160	0.66	0.30	0.04	80	2

	24	100	50	9	N/A			2.5	2.00	159	0.63	0.31	0.06
MEANS		**111**	**37**	**10**	**112**			**9.1**	**3.00**	**158**	**0.70**	**0.23**	**0.06**
PELLAND 1	1	98	23	7				14.5	4.26	128	0.77	0.18	0.05
2	2	91	22	8				12.9	4.14	121	0.75	0.18	0.07
3	3	77	23	6				12.1	3.35	106	0.73	0.22	0.06
4	4	89	25	11				10.8	3.56	113	0.71	0.20	0.09
5	5	96	22	11				12.4	4.36	125	0.74	0.17	0.09
6	6	N/A	18	7				N/A	N/A	N/A	N/A	N/A	N/A
7	7	85	22	8				13.2	3.86	115	0.74	0.19	0.07
8	8	76	21	7				11.7	3.62	104	0.73	0.20	0.07
MEANS		**86**	**22**	**8**				**12.5**	**3.89**	**116**	**0.74**	**0.19**	**0.07**
NON-CLOVIS CONTEXTS													
DUST CAVE 1	1	105	34	7					3.09	146	0.72	0.23	0.05
2	2	67	30	10					2.23	107	0.63	0.28	0.09
MEANS		**86**	**32**	**9**					**2.69**	**127**	**0.68**	**0.25**	**0.07**
BROOKEEN	113	105	43	12	110	13	8		2.44	160	0.66	0.27	0.08
	253	85	43	21	115	17	8		1.98	149	0.57	0.29	0.14
	236	102	33	9	N/A	7	4		3.09	144	0.71	0.23	0.06
	123	77	29	7	105	6	5		2.66	113	0.68	0.26	0.06
	74	97	37	6	120	5	3		2.62	140	0.69	0.26	0.04
	183	71	35	7	155	7	2		2.03	113	0.63	0.31	0.06
	77	64	36	7	125	11	4		1.78	107	0.60	0.34	0.07
	134	67	32	10	120	9	3		2.09	109	0.61	0.29	0.09
	120	91	32	9	125	12	5		2.84	132	0.69	0.24	0.07
	110	73	35	11	110	16	10		2.09	119	0.61	0.29	0.09
	97	69	32	7	120	10	5		2.16	108	0.64	0.30	0.06
	185	75	30	11	130	N/A	N/A		2.50	116	0.65	0.26	0.09
	107	74	34	4	120	N/A	N/A		2.18	112	0.66	0.30	0.04
	36	39	17	2	110	4	1		2.29	58	0.67	0.29	0.03
	162	68	34	7	115	14	5		2.00	109	0.62	0.31	0.06
	118	64	25	8	125	11	6		2.56	97	0.66	0.26	0.08
	104	42	19	4	115	8	3		2.21	65	0.65	0.29	0.06
	4	67	30	7	115	10	6		2.23	104	0.64	0.29	0.07
	78	82	36	9	115	19	9		2.28	127	0.65	0.28	0.07
	225	76	31	6	110	11	7		2.45	113	0.67	0.27	0.05
	115	87	51	12	N/A	N/A	N/A		1.71	150	0.58	0.34	0.08
	11	75	27	5	125	6	1		2.78	107	0.70	0.25	0.05

ART. #	MAX L	MAX W	MAX T	PLAT.>	WT g	PLAT W	PLAT D	INDEX CURV	RATIO W:L	L+W+T	L/L+W+T	W/L+W+T	T/L+W+T	APPROX. % CORTEX	STAGE
209	57	30	5	110		N/A	N/A		1.90	92	0.62	0.33	0.05		
10	74	29	8	120		N/A	N/A		2.55	111	0.67	0.26	0.07		
80	70	23	9	120		5	2		3.04	102	0.69	0.23	0.09		
164	79	30	7	N/A		N/A	N/A		2.63	116	0.68	0.26	0.06		
124	71	30	7	N/A		N/A	N/A		2.37	108	0.66	0.28	0.06		
651	103	47	18	125		N/A	N/A		2.19	168	0.61	0.28	0.11		
52	90	33	5	125		7	2		2.73	128	0.70	0.26	0.04		
235	78	30	9	120		7	3		2.60	117	0.67	0.26	0.08		
6	85	28	10	130		5	2		3.04	123	0.69	0.23	0.08		
108	69	26	9	N/A		N/A	N/A		2.65	104	0.66	0.25	0.09		
79	70	34	6	130		9	5		2.06	110	0.64	0.31	0.05		
135	77	26	8	110		4	2		2.96	111	0.69	0.23	0.07		
64	53	21	7	125		N/A	N/A		2.52	81	0.65	0.26	0.09		
231	84	30	5	125		8	3		2.80	119	0.71	0.25	0.04		
24	111	43	11	125		6	1		2.58	165	0.67	0.26	0.07		
50	102	39	10	130		9	3		2.62	151	0.68	0.26	0.07		
207	73	34	7	110		9	5		2.15	114	0.64	0.30	0.06		
MEANS	**77**	**32**	**8**	**120**		**9**	**4**		**2.41**	**117**	**0.66**	**0.27**	**0.07**		
GIBSON MEANS**	**85**	**42**	**14**						**2.02**	**141**	**0.60**	**0.30**	**0.10**		
WEAVER-RAMAGE MEANS**	**48**	**25**	**6**		**60**	**13**	**5**		**1.92**	**79**	**0.61**	**0.32**	**0.08**		
KIRCHMEYER	34	16	4						2.13	54	0.63	0.30	0.07		
	30	12	4						2.50	46	0.65	0.26	0.09		
	41	17	5						2.41	63	0.65	0.27	0.08		
	39	18	7						2.17	64	0.61	0.28	0.11		
	47	18	5						2.61	70	0.67	0.26	0.07		
	40	18	7						2.22	65	0.62	0.28	0.11		
	31	12	5						2.58	48	0.65	0.25	0.10		
	42	19	4						2.21	65	0.65	0.29	0.06		
	35	16	6						2.19	57	0.61	0.28	0.11		
	44	20	5						2.20	69	0.64	0.29	0.07		
	33	14	8						2.36	55	0.60	0.25	0.15		
MEANS	**38**	**16**	**5**						**2.31**	**60**	**0.63**	**0.27**	**0.09**		

INDIAN ISLAND							
55	18	6	3.06	79	0.70	0.23	0.08
50	19	7	2.63	76	0.66	0.25	0.09
60	23	9	2.61	92	0.65	0.25	0.10
52	21	6	2.48	79	0.66	0.27	0.08
47	24	6	1.96	77	0.61	0.31	0.08
46	18	7	2.56	71	0.65	0.25	0.10
47	22	7	2.14	76	0.62	0.29	0.09
34	13	2	2.62	49	0.69	0.27	0.04
34	15	3	2.27	52	0.65	0.29	0.06
36	13	3	2.77	52	0.69	0.25	0.06
45	20	12	2.25	77	0.58	0.26	0.16
33	15	16	2.20	64	0.52	0.23	0.25
40	13	3	3.08	56	0.71	0.23	0.05
40	15	5	2.67	60	0.67	0.25	0.08
43	14	7	3.07	64	0.67	0.22	0.11
42	17	6	2.47	65	0.65	0.26	0.09
MEANS 44	**18**	**7**	**2.51**	**68**	**0.65**	**0.26**	**0.10**
BARTON SITE							
85	32	22	2.66	139	0.61	0.23	0.16
62	24	10	2.58	96	0.65	0.25	0.10
79	16	8	4.94	103	0.77	0.16	0.08
88	22	10	4.00	120	0.73	0.18	0.08
110	34	10	3.24	154	0.71	0.22	0.06
33	14	5	2.36	52	0.63	0.27	0.10
48	20	7	2.40	75	0.64	0.27	0.09
MEANS 72	**23**	**10**	**3.12**	**106**	**0.68**	**0.22**	**0.10**
MUSTANG BRANCH**	70	37	12	1.89	119	0.59	0.31	0.10

* = ESTIMATED
** = PUBLISHED
AVERAGES

seems to be white with a surficial reddish staining, presumably from soil at its source. This chert is lustrous and opaque. Interior color is dark tan with thin bands—almost streaks—of lighter tan to almost white color. Concentric with the cortex on all three pieces is a band of color—dark gray on one piece, light gray on another, and grayish tan on the third. The dark gray band averages about 4 mm thick; the light gray, about 2.5 mm; and the grayish tan, about 2 mm. This material does not fluoresce under long- or short-wave ultraviolet light.

BLADES FROM, AND INFERRED TO BE FROM, THE CACHE

Metric data on the fourteen blades from the Keven Davis site are presented along with data on blades from other sites and caches in Table 6.1. As described above, it is possible to express shape of blades as a ratio of length to width to thickness, and to display this graphically on triangular coordinate diagrams (Fig. 6.4). This depiction of shape is achieved by summing length plus width plus thickness and then dividing that sum by each of the measurements and plotting the decimal proportions on the graph. By convention, length is greater than width and width is greater than thickness, so the plottings all fall in the same sector of the graph, and within that sector, general shapes are segregated as shown (Fig. 6.4).

The shapes of the nine complete Keven Davis blades are depicted on the accompanying triangular coordinate graph (Fig. 6.5). Simple width-to-length scattergrams visually display the two-dimensional shapes of blades (Fig. 6.6) and show the characteristically long and narrow form of the Keven Davis blades.

BLADE 1.
FIGS. 6.7 AND 6.8.
CHERT 1.

Blade 1 is in three pieces, lacks cortex, and has scars of five prior blade removals on its exterior, all removed in the same direction as this blade. None of these blade scars retains the negative bulb, indicating that a portion of the core had been removed in platform rejuvenation. There are three additional facets along the distal right margin that apparently remain from core preparation prior to blade removals. Moderate battering and crushing occur along a portion of the more prominent arris (Fig. 6.8 c). Lateral edges are nicked and chipped, particularly along the left margin (Fig. 6.8 a), which resembles Mallouf's category 4, common linear retouch; there are two complex edge nicks (Mallouf's category 2) on the right edge (Fig. 6.8 b). The distal edge is sharp and slightly nicked. The interior surface is very smooth, with low-amplitude ripple marks (Fig. 6.8 a) and a very slight bulb of percussion with one minute erraillure scar. There are several prominent and complex fissures on the interior near the distal end. Minor lipping occurs at the juncture of the interior and the platform. The platform is peaked and multifaceted without evidence of grinding. Over most of its length, the cross section of this blade resembles a trapezium. Near

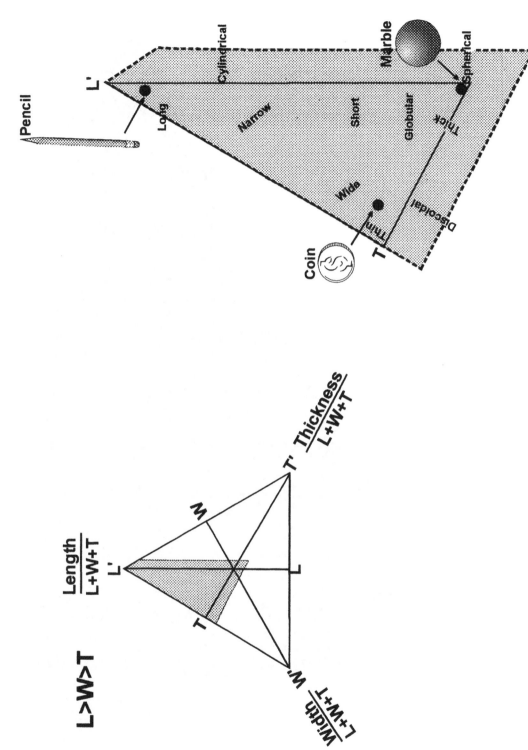

Figure 6.4. Triangular coordinate graph method of depicting shape (see text for explanation).

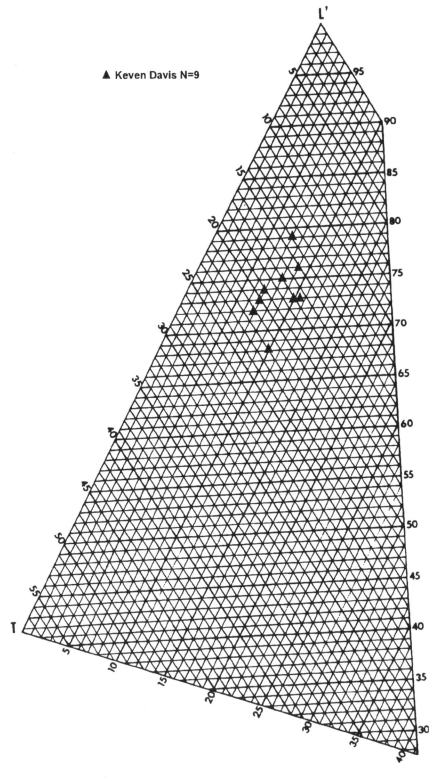

Figure 6.5. Triangular coordinate graphic presentation of the shapes of the nine most-complete Keven Davis blades. In this depiction, there is greater variation along the axis of length than along either of the other two primary axes; also, the generally long, narrow, and moderately thick shapes of these blades are seen to cluster fairly tightly.

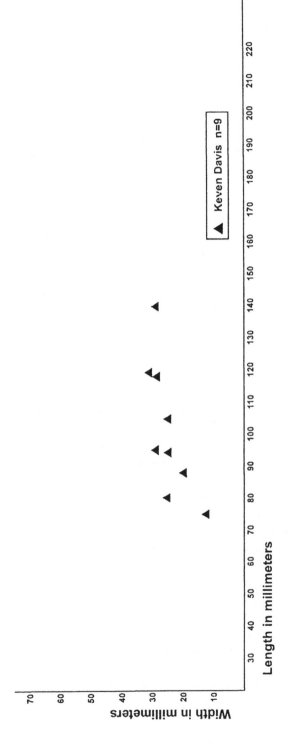

Figure 6.6. Scattergram of length (*x* axis) plotted against width (*y* axis) for the nine most-complete Keven Davis blades. The relationship is generally linear, with greater variation in lengths than in widths.

the distal end at the left margin is a clean bend fracture with an arris lip (Mallouf's category 11). Another bend fracture occurred 24 mm from the proximal end (Fig. 6.8 b). In both, bending was toward the exterior plane. This blade is in removal stage 6. The size and curvature of this blade precluded its microscopic examination by Kay. Two of the three pieces of this blade were found on, or exposed at, the surface near the center of grid unit 1, the position inferred to be the original cache spot (Figs. 5.1 and 5.2). The long fragment (115 mm long) was oriented nearly vertically in the ground, with the proximal fragment about 5 cm away to the southwest, lying on the surface. The small distal fragment was 47 cm southwest of the long fragment, 25 cm below surface.

BLADE 2.
FIGS. 6.7 AND 6.9 A.
CHERT 2.

Blade 2 was found in one piece. The exterior has one major scar from the prior removal of a blade and two smaller scars of what may have been core preparation flakes. The major scar is damaged at its proximal end but appears to retain a portion of its negative bulbar scar, indicating that it was removed from the same platform. There is cortex along the left margin, the distal portion of the right margin, and in a very narrow band across the distal exterior surface where the two lateral areas of cortex meet. Blade 2 is in removal stage 2. There is a small area of common linear retouch (Mallouf's category 4) at the distal end and along the right margin, where there are also one concave and one L-shaped complex edge nick (Mallouf's category 2) (Fig. 6.9 a). The platform is multifaceted and slightly dulled. A portion of the platform was removed by the L-shaped notch (Fig. 6.9 a). Most if not all of this edge modification is inferred to have been plow damaged. There is virtually no bulb on the proximal interior surface and very slight lipping at the juncture of the interior and platform surfaces. Ripples of moderate amplitude are present on the distal interior surface. There is one prominent fissure on the interior, somewhat proximal of the midpoint. Most of this blade has a triangular cross section. This piece was found in an almost vertical orientation near the surface in the inferred original cache spot.

BLADE 3.
FIG. 6.7, 6.9 B.
CHERT 1.

The distal portion of an incomplete blade, this highly curved fragment was found in one piece. There are parallel scars of three prior blade removals on the exterior. At the point of fracture, blade 3 is trapezoidal in cross section, but the central scar terminates short of the distal end of the blade where the cross section is triangular. The interior is very smooth, with low amplitude ripple marks. Both lateral edges are chipped and nicked with multiple, almost continuous flake scars (Mallouf's category 4—common linear retouch) (Fig. 6.9 b). Kay examined this piece microscopically (see Chapter 7) and found

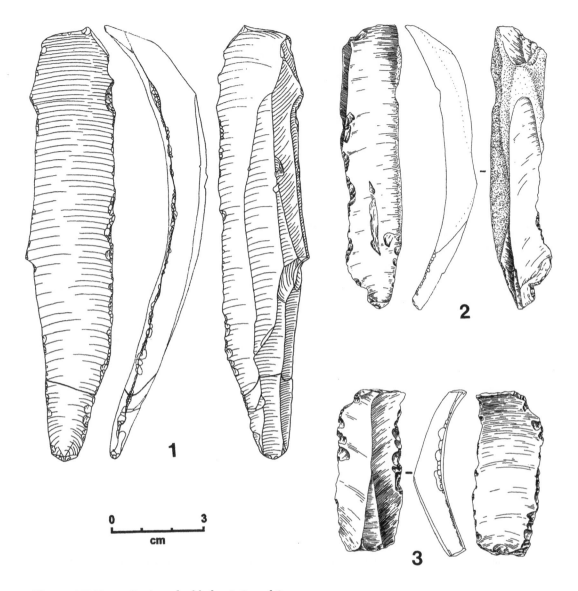

Figure 6.7. Keven Davis cache blades 1, 2, and 3.

evidence that the breakage and probably all of the edge damage were caused by machinery; no traces of aboriginal use wear were found. This fragment was found standing nearly vertically in the probable cache spot.

Blade 4 was found complete. The exterior of this small blade consists of one cortical and one blade-scar surface that meet in a ridge the length of the blade. Flake scars near the proximal end of the blade scar intrude into the area where evidence for any negative bulbar scar

BLADE 4. FIG. 6.10. CHERT 2.

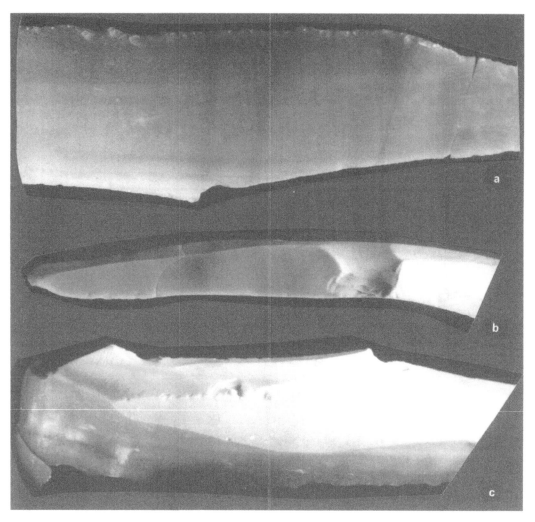

Figure 6.8. Details of damage seen on Keven Davis blade 1. Note (a) nicking and chipping of the lateral margin; (b) two complex edge nicks and a bend break; and (c) battering and bruising of the prominent arris.

should occur, but the remaining contours suggest that a negative bulbar scar had been present; this would indicate that no platform rejuvenation interceded between these two blade removals. The cross section is triangular. Both lateral edges are heavily modified with irregular but almost continuous flake scars (Mallouf's category 4, common linear retouch). The interior surface has a prominent bulb of percussion and moderately high amplitude ripple marks. The juncture of the interior with the platform is slightly lipped. The platform is peaked and shows light grinding. Blade 4 is in removal stage 2. This blade was found lying horizontally in loose soil very close to the inferred cache spot (Figs. 5.1 and 5.2).

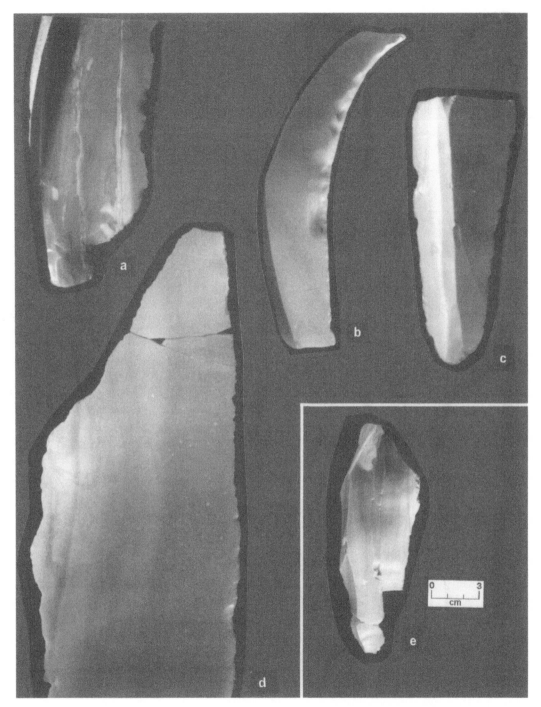

Figure 6.9. Details of Keven Davis blades 2, 3, 5, 6 and refit of blades 6 and 14. Note (a) complex edge nicks along right margin of blade 2; (b) common linear retouch along the right margin of blade 3; (c) arris hinge scar adjacent to the transverse fracture plane on blade 5; (d) simple lateral wedge snap at the distal end of blade 6; and (e) refit of sequential blades 6 and 14.

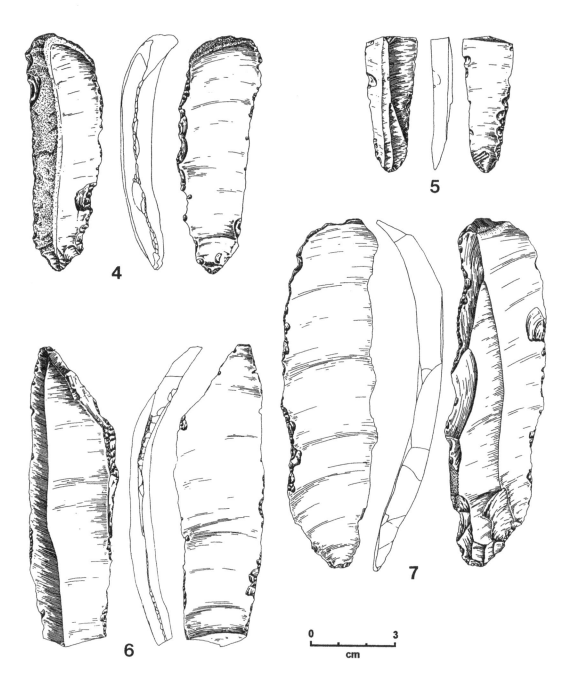

Figure 6.10. Keven Davis blades 4, 5, 6, and 7.

Blade 5 is a proximal blade segment. The exterior surface exhibits scars of three prior blade removals and one prior flake removal in the same direction as this blade. Partial remnants of negative bulbar scars are present on two of the blade scars. No cortex is present. Both edges have common linear retouch (Mallouf's category 4) scars on the exterior surface and there is an arris hinge scar (Mallouf's category 11) in conjunction with the snap fracture of the blade (Fig. 6.9 c). The present segment lacks longitudinal curvature, although it is likely that distally it was curved. The interior is highly smooth, with negligible ripple marks; there is a very slight bulb with two small erraillure scars. Some of the common linear retouch invades the interior surface. The cross section resembles a trapezium. A very minor lip occurs at the juncture of the interior and platform surfaces. The platform is multifaceted and ground. I have placed this in removal stage 6. Kay, in his microscopic examination of this piece, was able to determine conclusively that it was broken and scarred by contact with a metal plow. Blade 5 was found 10 cm below the surface about 20 cm southwest of the inferred cache position.

BLADE 5. FIGS. 6.9 C AND 6.10. CHERT 1.

Blade 6 is approximately the distal half of a large blade and was found in three pieces. The exterior has two prominent, parallel blade scars removed in the same direction. Additionally, there are two minor flake scars oriented transversely and a small area of cortex near the distal right margin. The scar on the left side of the arris on blade 6 refits with the interior scar on blade 14 (Fig. 6.9 e); the scar on the right side of the arris on blade 6 is continuous with a scar on the right exterior margin of blade 14. Blade 6 was removed after blade 14 without any intervening removals from this area of the core face. Common linear retouch (Mallouf's category 4) occurs along both margins, with scars invading both faces. The interior is smooth, with low-amplitude ripple marks. There are numerous, minute fissures on the interior. The cross section is triangular. Blade 6 is in removal stage 5. The fractures near the distal end (Fig. 6.9 d) constitute a simple lateral wedge snap (Mallouf's category 6). The three fragments of which this blade segment has been reconstructed were found in two locations. Pieces 7 and 8, the wedge and small distal fragment, were found a few millimeters apart, 10 cm below the surface at the northeastern edge of the inferred position of the cache. The larger piece, fragment 10, was found southwest of these at a distance of 35 cm, oriented vertically, from 10.5 to 19 cm below surface.

BLADE 6. FIGS. 6.9 D, E AND 6.10. CHERT 1.

Blade 7 is an intact blade. The exterior has three major planes, two formed by prior blade removals in the same direction as this blade and one composed of cortex and exhibiting an oblique flake scar from early core preparation. Several small flake scars on the proximal, exterior

BLADE 7. FIGS. 6.10 AND 6.11 D. CHERT 1.

surface of this blade preclude any observation on the presence of negative bulbar scars. The right lateral edge has one simple edge nick and discontinuous common linear retouch (Mallouf's categories 1 and 4, respectively). On the distal edge there are two small simple edge nicks (Mallouf's category 1). A continuous series of scalar flake scars occurs along 48 mm of the left lateral edge from about midpoint to near the distal end. These are difficult to diagnose and could be either intentional retouch or, more likely, damage—Mallouf's category 4, common linear retouch (Fig. 6.11 d). This blade was recovered in a large soil ped (Fig. 6.12) that has been preserved intact and has a perfect impression of a portion of the left margin of the blade. This circumstance of discovery clearly indicates that the blade edge was modified long before the 1988 removal of fill from this location—at least long enough for this soil ped to form around the piece of chert. This is consistent with the observation that edge modification and breakage among the pieces of this cache are not discernibly fresh in appearance when compared to blade surfaces, whereas damage that occurred during archeological excavation (for example, to blade 14) is perceptibly fresher in appearance. The interior of blade 7 is smooth, with no bulb and very low amplitude ripple marks. There are a few very minor fissures. In cross section, blade 7 approximates a trapezoid. The platform, which is multifaceted and lightly ground, intersects with the interior surface with a minor lip. This blade is in removal stage 4 and is only moderately curved. Blade 7 was found 30 cm southwest of the inferred cache spot, lying at a slight inclination, 12 to 16 cm below surface.

BLADE 8.
FIGS. 6.11 B, C
AND 6.13.
CHERT 1.

Now in two pieces, blade 8 was complete when cached. The exterior has a small area of cortex along the right margin near the distal end and scars of three prior blade removals in the same direction as this blade. One of these blade scars seems to retain a negative bulbar scar, one definitely lacks this feature, and the proximal surface of the third is absent. There are minor flake scars on the exterior along the edges, part of common linear retouch (Mallouf's category 4). The blade is broken by pressure snap with dorsal scaling (Mallouf's category 12) near its midpoint (Fig. 6.11 c). This fracture transects an edge spall (Mallouf's category 3) along the left margin of the blade (Fig. 6.11 c). A minute wedge is missing from the fracture at the right margin, possibly indicating that this fracture matches Mallouf's simple lateral wedge snap (category 6). The blade interior is smooth, with low-amplitude ripple marks, a very slight bulb, and neither fissures nor erraillures. In most of its length, the cross section of this blade resembles a trapezium. A minute lip is formed at the juncture of the interior surface with the platform, which is single faceted. This blade is in removal stage 4 and strongly curved. Kay examined this piece

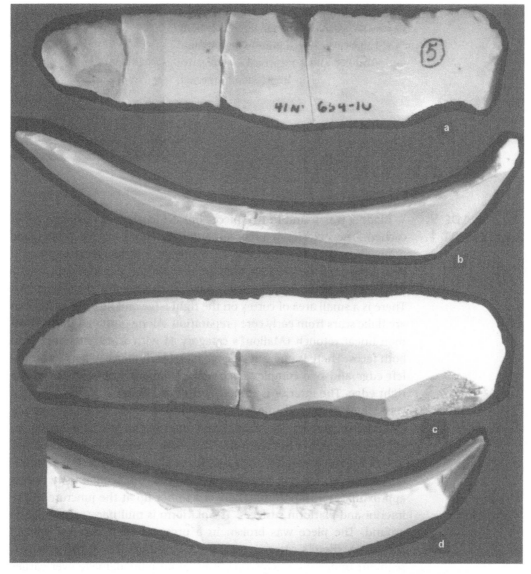

Figure 6.11. Details of Keven Davis blades 7, 8, and 10. Note nicking and chipping of lateral margins of (d) blade 7; and edge chipping, dorsal scaling, edge spalling, and simple lateral wedge snapping forms of damage to (b, c) blade 8 and (a) to blade 10.

microscopically, particularly in the area of the transverse break, and determined that the breakage and associated edge modifications all occurred as plow damage. The two fragments from which this blade was reconstructed were found 50 cm (distal fragment) and 100 cm (proximal fragment) southwest of the original cache position, at depths of 25 and 33 cm, respectively.

Figure 6.12. Large soil ped with imprint of Keven Davis blade 7; marginal retouch along the right margin occurred prior to formation of this ped and is preserved as a negative impression in the soil.

BLADE 9.
FIG. 6.13. CHERT 1.

Blade 9 was found in two pieces but evidently was complete when cached. The exterior is predominantly the scars of two prior blade removals in the same direction as this blade. For one, there appears to be a portion of the negative bulb present and for the other, the proximal portion of the scar is not present, precluding any determination. There is a small area of cortex on the right edge and near the distal end are flake scars from early core preparation. Along both margins is common linear retouch (Mallouf's category 4) with scars extending onto both faces. There is one simple edge nick (Mallouf's category 1) on the left edge and two complex edge nicks (Mallouf's category 2) on the right edge. The interior is smooth with low amplitude ripple marks, a very slight bulb, and minor fissures. Strong common linear retouch (Mallouf's category 4) invades the interior surface and forms an oblique edge left of the platform. This blade is moderately curved and fits into removal stage 6. Over roughly half of its length, it is triangular in cross section, but near both extremities, this section more closely approximates a trapezium. There is a minor lip at the juncture of the interior and platform surfaces. The platform is multifaceted and lightly ground. The piece was broken in a pressure snap and has ventral scaling (Mallouf's category 12). In his examination of the proximal segment under the microscope, Kay found plow damage over earlier striations which he interprets as probable use wear from a sawing/ cutting action. This would indicate that this blade was cached intact after having been used, but apparently not worn out. These two pieces were found 65 cm (distal fragment) and 75 cm (proximal fragment) southwest of the inferred cache spot at 23 and 18 cm below surface, respectively.

BLADE 10.
FIGS. 6.11 A
AND 6.13.
CHERT 1.

Blade 10 was found in 4 conjoinable pieces, clearly broken by machinery. The exterior of the reconstructed blade has the scars of three prior blade removals in the same direction as this blade, as well as two core-preparation flake scars that are near the distal end on the

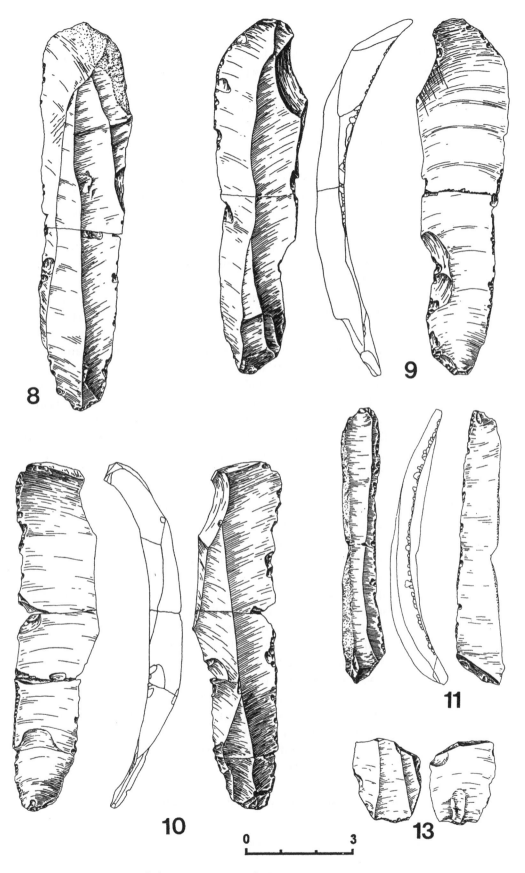

Figure 6.13. Keven Davis Blades 8, 9, 10, 11, and 13.

left side. One negative blade facet seems to retain a portion of the bulbar scar, and the other two are indeterminate. There is discontinuous common linear retouch (Mallouf, category 4) along both margins. On the interior are observed a smooth surface, no bulb, low-amplitude ripple marks, and numerous small fissures. The cross section is triangular except near the proximal end, where it is trapezoidal. The juncture between the interior and the platform is lipped. The platform is single faceted and lightly ground. The blade is strongly curved and belongs to removal stage 6. The three fractures (Fig. 6.11 a) are typical of plow damage (Mallouf's categories 7 and 12). The more-distal fracture is the most complex, being a complex lateral wedge snap with arris damage and retouch of the fracture planes of both blade segments. The middle break is also a complex lateral wedge snap with arris damage and retouch of the fracture surface of only the segment distal to the break. The proximal break is of the pressure snap variety with a single large lip scar on the interior surface (Fig. 6.11 a). Microscopic examination of this piece revealed extensive metallic contact and clear plow damage, and also an earlier, weathered set of randomly oriented striae. Kay interprets these as indicating damage from transporting this blade (or its parent core prior to detachment of this blade). The four pieces from which blade 10 was reconstructed were found dispersed in the excavation. One fragment was in the inferred cache spot, 7 cm below the surface. Three were about 10 cm apart some 80 cm southwest of the first at 15, 22, and 29 cm below the surface.

BLADE 11.
FIGS. 6.13 AND
6.14 A, B.
CHERT 2.

Blade 11, a small specimen, was found in two fragments. Also, a small chip recovered by screening refits onto the two fragments at the point where they conjoin. The blade's exterior surface has two prior blade scars removed in the same direction as this blade, one of which retains its negative bulbar scar. There is also a cortical facet extending almost the entire length of the blade along its right margin. The nick along the right margin (where the recovered chip refits) and discontinuous retouch along the left margin (Fig. 6.14 a, b) could be interpreted in one of two ways. In one view, these might be considered common linear retouch (Mallouf's category 4) along the left margin and a simple edge nick (Mallouf's category 1) on the right margin; this nick is coincident with a simple snap fracture through the center of the blade—a combination not reported as occurring in plow damage by Mallouf (1982) but not necessarily precluded. It appears that if force were exerted on the interior surface, near the right margin of this blade, it would cause both the snap and the removal of the chip. Alternatively, since no microscopic evidence of contact with the plow was found and there was good microscopic evidence for use wear (see

Chapter 7), it could be inferred that this blade was used and was intentionally snapped just prior to being cached, and that the detached chip was then also included in the cache.

In view of the other plow breakage, it seems more likely that this blade was cached intact and broken by plowing. That no metallic streaking is present might indicate that the force of the plow was exerted with enough intervening soil matrix to protect the chert from full contact with metal. This interpretation leads to the conclusion that the blade was complete when it was used and when it was cached.

The distal terminus of this blade is difficult to interpret. It consists of a snap fracture that caused a small piece to be broken from the arris. These fracture planes are equally weathered with the blade surfaces, and it would appear that the end was detached prehistorically. The interior is smooth and lacks bulb, ripple marks, and fissures, but there is a small erraillure scar. The cross section of this blade approximates a trapezium. There is no lip at the juncture of the platform and the interior surface. The platform is multifaceted and ground. The blade is curved and belongs to removal stage 2. The two pieces making up this blade were found 85 and 100 cm southwest of the inferred cache spot, at 31 and 19 cm below the surface, respectively.

**BLADE 14.
FIGS. 6.14 C, D, E
AND 6.15. CHERT 1.**

(Note: the small piece formerly designated blade 12 was found to refit with blade 1 and has already been described; blade 13 is described in the next section.) Blade 14 was the largest in this cache, measuring over 14 cm long even with a small portion of the proximal end missing. The blade's exterior retains four core-preparation flake scars along the left margin and the scars from two prior blade removals, both of which appear to have originated from a prior platform. There is battering of the arris near the midpoint of the blade, probably resulting from an effort to remove an irregularity on the arris prior to detaching this blade (Fig. 6.14 d). Toward the distal end of this blade, there is an area of subcortex. Discontinuous common linear retouch (Mallouf, category 4) is present along most of the left margin and in limited places along the right margin. The interior is smooth, with almost no ripple marks and no fissures. In cross section over most of its length, this blade resembles a trapezium. This blade belongs to removal stage 5. There are three breaks and one significant nicking of this blade, occurring at two or three different times. Near the distal end is an incomplete complex lateral wedge snap (approaching Mallouf's category 7). This has all of the features of Mallouf's category 7 except that the wedge did not fully detach, but a hairline fracture defining it can be seen on the interior of the blade (Fig. 6.14 e). Otherwise, the features are present—arris damage, retouch of the single fracture plane that did detach, and the complex edge nick on the margin of the blade.

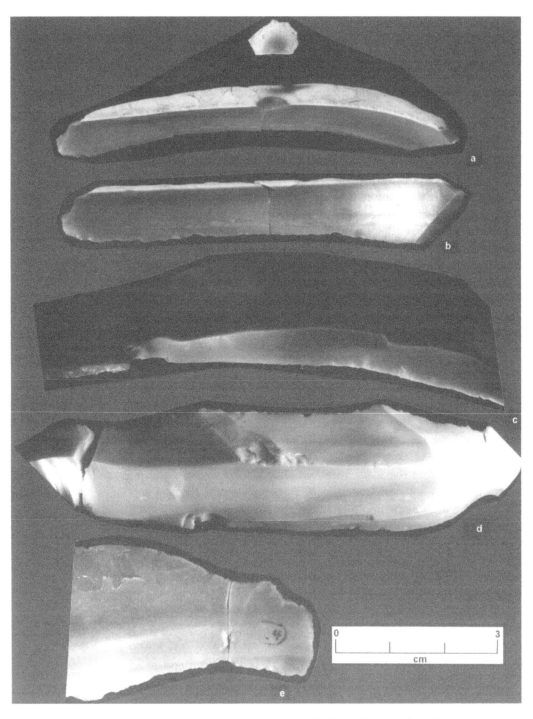

Figure 6.14. Details of damage on Keven Davis blades (a, b) 11 and (c, d, e) 14. Note (a) on blade 11 the combined snap and nick fractures near the center of the blade; and (c), on blade 14, marginal retouch along the margin, (d) battering of the arris, and (e) incompletely detached wedge snap.

It is inferred that this fracture occurred as the result of plow damage in the recent past, the two pieces being found approximately 75 cm apart. The fragment distal to the just-described break is about 1.1 cm long; the one proximal, about 13.3 cm long. This longer fragment manifests the damage next described. The proximal end of the blade is missing, and the weathered fracture scar indicates that a lipped bend fracture occurred, but it is not clear when this may have occurred (perhaps prehistorically, perhaps in plowing in the recent past). The remaining two fractures occurred during archeological excavation and are fresher than any of the other surfaces on this piece. In both cases the edge of the blade was struck with a metal excavation tool (Young, personal communication, 1989). The first, near the proximal end of the blade, nicked the right margin and created a transverse break with arris damage and retouch of the fracture surfaces, resembling Mallouf's complex lateral wedge snap (category 7). The second, 32 mm down the same edge, consists of a complex nick and an edge spall 46 mm long (Fig. 6.14 c). The two conjoined segments of blade 14—disregarding the contiguous pieces broken during excavation—were found about 175 cm southwest of the inferred cache spot, 25 cm below the surface.

The two pieces of blade 15 were found in 1991 in the housing development area where fill dirt borrowed from the cache area had been spread. In every observable attribute, this reconstructed blade matches the blades already described, and it seems a virtual certainty that it had originally been a part of the cache. The exterior has five facets of blades removed in the same direction as this blade. There are no negative bulbar scars, indicating that platform rejuvenation had occurred. There is a small remnant of subcortex near the distal terminus. Both margins exhibit common linear retouch (Mallouf's category 4), and there is one edge spall (Mallouf's category 3) adjacent to the break and another on the proximal right margin. The interior is smooth, with moderate-amplitude ripple marks, a slight bulb, and a few minor fissures. In cross section over most of its length, this blade is a five-sided prism with four exterior facets and one interior facet. The platform is missing, evidently the result of crushing at the time of blade removal. This blade is in removal stage 6. The break that occurs near the midpoint is a pressure snap (Mallouf's category 12) with damage at the arris and scaling on the blade interior. Kay's microscopic examination revealed metallic streaking on both faces, implicating plowing as the cause of the breakage and edge modification of this blade. As mentioned, the two pieces of this blade were found by Keven Davis in 1991 where fill dirt had been spread in Lakeview Estates Subdivision.

It seems clear that unrecovered fragments of the three incomplete

BLADE **15**. FIG. **6.15**. CHERT **1**.

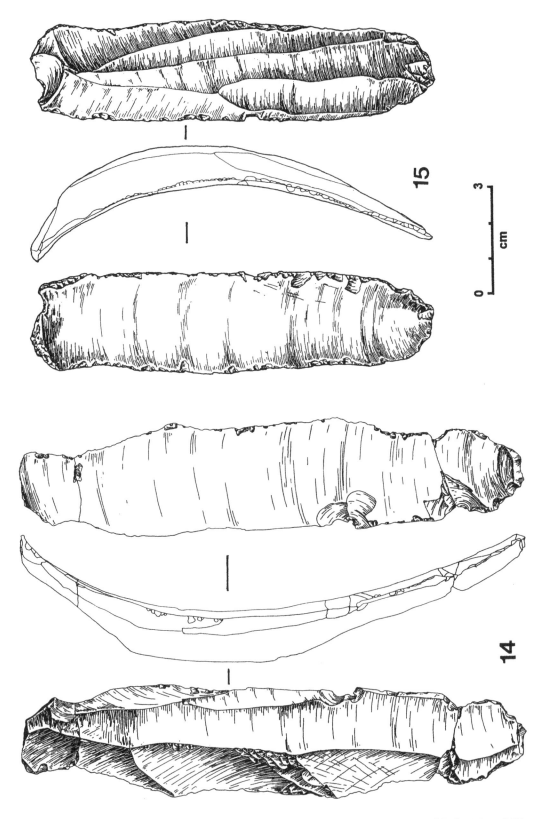

Figure 6.15. Keven Davis blades 14 and 15.

blades (3, 5, and 6) and possibly an unknown number of other blades or blade segments might have been dispersed with the fill borrowed from the cache site.

One proximal blade segment (blade 13) from the locality is distinct from those just described in that it is completely patinated, including the fracture plane and edge damage. It was found in a disturbed area 75 meters west of the boat ramp (Fig. 4.3). Fill dirt had been spread in places in this general area, but it is unclear from the site records whether this piece seemed to have been transported and dropped here along with fill from the lake-bed borrow area. Because this piece is patinated, it seems unlikely that it has experienced the same life history as the other blades from the site. Two alternative interpretations suggest themselves. First, the piece may never have been cached with the other blades and lay for a long period of time on the surface where it was found. Second, it may have been part of the cache until a disturbance of some kind brought it to the surface where it lay exposed and became patinated. Subsequently, when borrowing occurred, this already patinated surface piece may have been relocated along with a load of fill dirt. Since it is not known how long this chert would have to be exposed in this environment to become this patinated, it is not known when such a disturbance may have occurred. If patination were rapid, land clearing or plowing earlier in this century are possibilities; if it were slow, perhaps an earlier, natural disturbance, such as the burrowing of an animal, was responsible. Given the lack of general archeological debris at this entire locality, it would seem more likely than not that this piece was once part of the cache, and it is here described, as follows.

BLADE FRAGMENT AND LARGE CHERT NODULE

A deeply patinated proximal blade fragment slightly more than 2 cm long is designated blade 13. The exterior has four facets parallel to the longitudinal axis. The central facet has most of the negative bulbar scar of an earlier blade, a feature not seen as clearly on any of the other blades from this locality. Each of the margins has a single simple edge nick (Mallouf's category 1). The interior surface is smooth and has a moderately prominent bulb with a large erraillure scar. The cross section at the point of breakage is prismatic, with four exterior and one interior facets. The interior surface intersects with the platform with a well expressed lip. The platform is multifaceted and lightly ground. Its configuration is recurvate in two planes (Fig. 3.19 a), like the well-known *chapeau de gendarme* of Levallois point typology (cf. Bordes 1953; Fish 1979). This is the one blade from the site that seems likely to have been detached using direct, soft-hammer percussion. It is inferred that this blade would conform to removal stage 6,

BLADE 13. FIG. 6.13. CHERT 1.

although, since the distal segment is missing, this cannot be a firm inference. The fracture plane at the distal end of this segment is a pressure snap (Mallouf, category 12) that is somewhat crenelate and is continuous, with a nick on the interior surface adjacent to the fracture plane. Since the three (two on the blade edge and one on the fracture edge) simple edge nicks and the nature of the fracture plane conform in detail with damage seen from plowing (Mallouf 1982) and are patinated equally with the original blade surfaces, I favor the interpretation that this piece was probably broken and brought to the surface by plowing earlier in this century, patinated, and was dispersed with fill dirt in the recent borrowing.

Keven Davis also found a large chert nodule (Fig. 6.16) some two kilometers from the cache site. This was a surface find and, like the cache, did not appear to be in a site with other debris or features. This nodule is remarkable in several respects. First, it is of Chert 1, identical to that of which the majority of the cache blades are made. Second, it is very large (26 x 23 x 20 cm, maximum dimensions) and weighs 18 kilograms. Third, it exhibits limited flaking suggestive of the early stages of preparing a polyhedral core. The natural shape of this piece was evidently ovoid. Most of its exterior is covered in soft, creamy white cortex, but one end has been flaked off obliquely to the long axis of the nodule, forming a faceted surface similar to that of Clovis blade-core platforms. Since this is a surface find, it cannot be unequivocally tied to any particular time period or archeological manifestation. However, the affinities with the cached blades in terms of location, type of chert, and manner of flaking—as if to set up a blade core platform—raises the possibility of historical relatedness. This possibility is more intriguing when considered in light of the relatively infrequent occurrence of debitage of Edwards chert in Archaic and Late Prehistoric sites in the area (McGregor 1987). Mallouf (1994:45–46) infers on technological grounds that the Sailor-Helton cache found in southwestern Kansas is of Clovis affinity. He notes that the raw material is entirely Alibates agate, which originated some 200 kilometers to the south. Total weight of the 10 cores, 151 blades and flakes, and 5 unifaces in the Sailor-Helton cache is 20.53 kilograms. In both its weight (18 kilograms) and the distance it was transported (ca. 200 kilometers), the large nodule found by Keven Davis compares closely with the Sailor-Helton cache. Even though the alternative explanation of pure coincidence seems remote, the possibility that this large nodule was brought to the Navarro County area by Clovis knappers cannot be substantiated given presently available information.

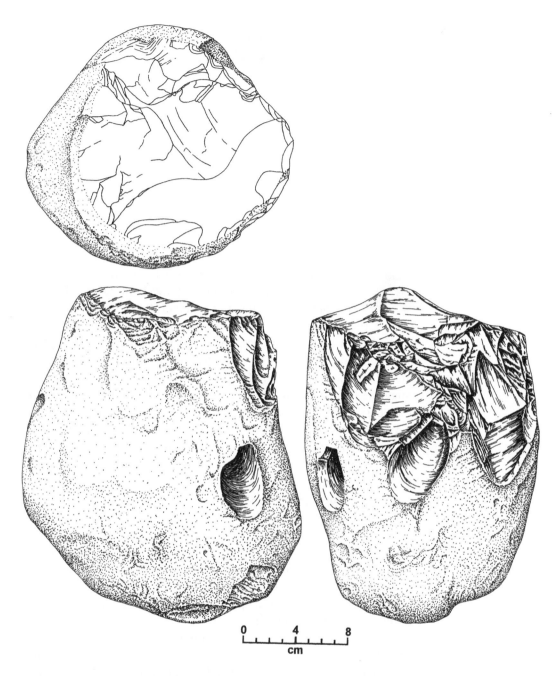

Figure 6.16. Very large nodule of Edwards chert, Georgetown variety, found in Navarro County near the Keven Davis site. Flaking has removed one end of the nodule.

SEVEN *Microscopic Attributes of the Keven Davis Blades*

MARVIN KAY

INTRODUCTION

A problem with the Keven Davis blade cache, related to its discovery's being occasioned by mechanical stripping, was well stated by Young and Collins (1989: 26): "Fresh breakage [of the twenty-three recovered specimens] is not easily distinguished from original surfaces, and . . . it is not certain which pieces are intentional segments as distinct from machine damage." They (Young and Collins 1989:27) further observed that lateral edge flaking "suggests damage from vigorous use rather than intentional retouch. It appears that edge damage from machinery is minimal."

As a consequence, I undertook a microwear analysis of the cache at the request of Collins. My analysis follows the approach developed by S. A. Semenov (1964), whose primary impact has been in Russia and the former republics of the Soviet Union, where "traceological" studies of microscopic polishes, striations, and edge damage are used to delineate tool function, or use. Empirically grounded in thousands of experimental tool-use replications carried out under naturalistic field conditions by either Semenov or his successor, G. F. Korobkova, and her staff, the traceological approach has placed priority in the analysis of striations on tool edges and surfaces as the key to understanding tool use and, to a lesser degree, materials worked (Phillips 1988; Korobkova, personal communication, 1991).

Striae and other evidence of abrasion are, indeed, among the more compelling microwear traces. Echoing comments Semenov and other traceologists have made, Cotterell and Kamminga (1990: 159) correctly observe, "The importance of use-scratching . . . is that it provides evidence of how a tool was oriented during use as well as about the nature of the material worked." Post-depositional movement or other taphonomic processes can also cause or substantially alter lithic microwear and would be a measure of disturbance, displacement, or the environment or climate at or after burial. Thus, as we indeed shall show, a wear-trace analysis is likely to help resolve three prime questions originally anticipated by Young and Collins: What caused the breakage of the prismatic blades? What caused the edge damage? And what inferences of tool use are possible?

SAMPLE CONSIDERATIONS

Of the total of fourteen individual whole or fragmentary prismatic blades, five specimens (blades 1, 2, 6, 7, and 14) were excluded because they are generally too thick or have longitudinal section curvature that is excessive for proper mounting for microscopic examination. Once unglued, however, fragments of nine blades, one whole specimen, and one flake from an unidentified blade were subjected to an intensive, systematic microwear examination at intermediate magnifications (100–400X) that largely resolved the three research questions. The selection of specimens was made solely on the basis of object size and shape, not on the provenance of the finds. The analyzed sample includes specimens from both surface and excavated, subsurface contexts, as well as the one specimen (Fig. 7.1 c) found a substantial distance (ca. 150 meters) away from the inferred cache spot.

Microwear analogs from experiments in the processing of soft plant materials, wood, bone, or antler and from the butchering of game or the effects of stream tumbling using replicated chipped-stone artifacts, as well as other archaeologically obtained artifacts, also were compared with the microwear from the Keven Davis cache. These ongoing, comparative studies are partially described elsewhere (Frison 1986, 1989; Kay 1996, 1997, n.d.), and their results are selectively incorporated here.

METHODOLOGY

The microwear approach, which allows for sequential macroscopic and microscopic observations, is patterned after the work of Lawrence H. Keeley—who virtually alone reinvigorated microwear research in western Europe and America—but eliminates the two-dimensional image problem he faced (Keeley 1980: 12–14). Although lower magnifications are employed, most helpful is the intermediate magnification range of 100 to 400 diameters using a reflected-light differential-interference microscope with polarized light Nomarski optics (Hoffman and Gross 1970). The optical qualities of this microscope are superior to that of most incident light microscopes presently used for microwear studies, but—to my knowledge—have received only minimal attention by microwear analysts in western Europe and America (see Dumont 1982 for a quantified application using a similar microscopy system). The Nomarski optics capability of this system is ideally suited to lithic microwear analysis because its color divisions of polarized light allow for three-dimensional views of tool surfaces. Image resolution, or clarity, increases with magnification (in contrast

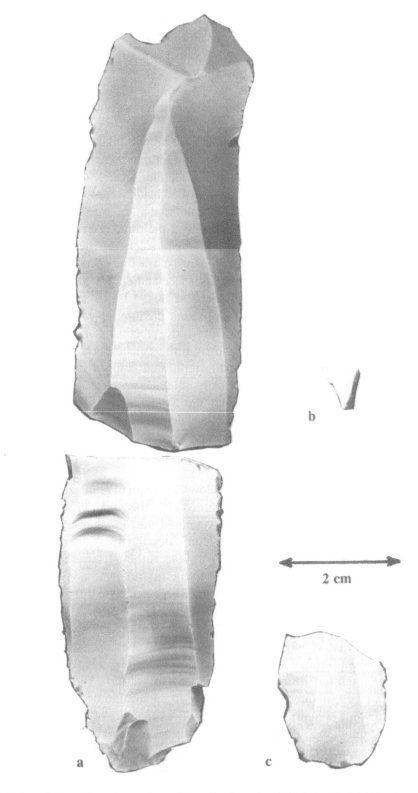

Figure 7.1. Dorsal view of specimens from Keven Davis cache: (a) blade 15; (b) flake from water-screening residue; and (c) blade 13. All specimens have been "smoked" with ammonium chloride for photographic purposes only.

to incident light microscopes but similar to scanning electron microscopes, while depth of field decreases. Unlike scanning electron microscopy, this system does not require the metallic coating of tool surfaces. Striations and micropolishes are readily observed, and easy adjustment of the Nomarski optics allows for the best display of wear traces. Micropolishes are optically bright and characteristically undifferentiated. Thus, polish brightness—regarded by Keeley (1980: 23–24, 62–63; see also the comments by Bamforth 1988: 13) as the prime signature of different contact materials and the measure to distinguish among different polishes, even on surfaces of different textures—is not used as a criterion to identify contact materials. For most archeological specimens, it is also unnecessary to carry out the extensive and potentially destructive chemical cleaning that Keeley (1980: 11) advocates to ensure that the microscopic inspection of a tool surface or edge is unimpeded by organic or inorganic residues.

Some experimental tool replicates were, however, chemically cleaned in accordance with Keeley's general approach. The major difference was in substituting KaOH for NaOH to remove organic residues and to reduce the likelihood of chemically patinating the tool surfaces, a problem with NaOH (Keeley 1980: 11) but not KaOH. The Keven Davis archeological specimens were not subjected to any chemical cleaning beyond the occasional use of methyl alcohol to remove oily films from tool surfaces. These came from the plastocene clay used to mount an artifact on a slide plate or from previous handling prior to my analysis. Latex surgical gloves were worn while handling both the chemically cleaned and uncleaned specimens, except when lateral edges were felt for edge irregularities and grinding. The Keven Davis blades, however, were almost invariably broken, and conjoinable fragments had been reattached with a commercial, water-soluble glue prior to this study. To loosen the glue bonds, examined specimens were first placed in individual clear, clean polyethelene plastic bags filled with a water and mild, liquid hand-soap solution, and then ultrasonically agitated. After about thirty minutes or less, the ultrasonic cleaning broke the bonds without further damaging the conjoinable fragments.

At magnifications greater than 100 diameters, wear traces can be distinguished from edge damage and overall flaking characteristics. Indeed, at these magnifications, any concern about reliably separating edge damage from intentional bifacial retouch (Frison 1979: 264–267) becomes irrelevant to the technofunctional analysis of tool wear. Bifacially flaked stone tools can be evaluated for microwear traces in the same manner as unifacially flaked or unretouched stone tools because, at this higher magnification, wear traces crosscut or extend beyond the grosser details of microflaking. In either case, oriented micropolishes and striae can be located, photomicrographed, and measured as needed.

In evaluating wear traces, the full range of magnifications was employed, going from 100 to 200 and then 400 diameters and using both bright and dark field illumination. Artifact surface scans of both surfaces and the edges were accomplished at magnifications of 100 or even 200 diameters. Further examinations at 400 diameters were done whenever possible to better characterize wear traces.

Photomicrographs were routinely taken with a 4 x 5 in. camera back and Polaroid Professional 55 positive/negative film. A green filter and dark field illumination were used in many cases to improve, or enhance, the contrast of the photomicrographs. The orientation and location of the photomicrograph were recorded relative to that of the artifact. Photomicrograph size was determined by photographing a micrometer disk scale at the appropriate magnifications. Wear traces were further described, even when not photomicrographed. The location of wear traces relative to edge damage discovered at intermediate-range magnifications also was noted in the comparisons between macroscopic and lower-powered magnifications. In most cases there is a direct correlation between edge damage and wear traces near or at an artifact edge.

Minor but important additions to the overall methodology also have been made. Subsequent to the examination of the experimental tool replicates, the orientation data for photomicrographs have been systematized. Initially, I noted (and continue to note) the photomicrograph orientation on a contact print with respect to which end of the photomicrograph faced the proximal or distal end of the artifact. I now routinely position a photomicrographed artifact still mounted on its microscope slide plate onto either an enlarged photograph or outline drawing of the artifact and then inscribe the edge of the slide plate onto the illustration; by so doing, it is possible to get a reasonable match with the actual photomicrograph orientation. I also, on occasion, take overlapping photomicrographs to further show the arrangement of wear traces within a photomosaic. Because slight changes in the dip of the object can dramatically change the microscopic image, it is also useful and instructive to reposition an artifact or tool replicate on a slide plate and then reexamine it for wear traces.

The microwear patterns are independent of stone materials (Kay 1996, 1997) and are observable at magnifications greater than 100 diameters. The overall methodology complements and provides, in my estimation, more specific information about tool use than is available through either macroscopic or low-power magnification (<100X) microwear studies alone. It also deals successfully with artifacts that normally are excluded from microwear examination—

namely, bifacially flaked tools whose formalized shapes are functionally subdivided. The analyses described here, however, take substantially longer to perform and also require more expensive technical equipment than that needed for either macroscopic or low-power magnification microscopic tool-use evaluations. The added expense in time and equipment, however, is justified by the results.

In the present study the most crucial information is functionally diagnostic of general classes of stone tools, of contact materials, and of the geomorphic or archeological contexts. Specific analytical concerns are: (1) the placement, orientation, and cross-cutting relationships among use-wear striations, plus their width, depth, and number; (2) the presence of abrasive microparticles in relation to striae; (3) for polishes (which in this case clearly formed through frictional processes), the degree of development, texture, area, and placement on a tool surface or edge; (4) for edge damage, its relationship to other use-wear traces; and (5) identification of inorganic and metallic residues.

MICROWEAR OBSERVATIONS

The cause of blade breakage is of singular interest, and much of the microwear analysis was directed to resolving the issue. This task turned out to be relatively easy, as the microwear evidence associated directly with breakage is clear and convincing. Blade 8 is instructive of the main lines of evidence and is described in detail.

Blade 8 is broken into two pieces (both recovered in the subsurface excavations) by a fracture transverse to the longitudinal axis, a relatively common breakage pattern for the cache. Secondary flaking associated with this fracture is prominent on the ventral surface and edges of the two pieces. The proximal fragment, when viewed from the ventral surface, has a "burinlike" facet on the right edge of the break that could have occurred coincidental to the breakage or that might represent intentional modification for usage as a burin. There is also a little flake scar on the ventral surface at the point of the "burin facet" at the break.

Microscopic evaluation of the transverse fracture, the "burin facet," and the surfaces and edges of the two blade fragments demonstrate that breakage occurred as the byproduct of mechanical damage. Figure 7.2 b shows the distal portion of the "burin facet." The presence of a metallic residue adhering to its surface and occurring as a streak originating at the facet edge and running parallel to the longitudinal axis of the blade on the ventral surface conclusively shows the cause of the break and the cause of the "burin facet" are one and the same. A complementary metallic residue is also present on the opposite edge (Fig. 7.2 a) of this fragment, and other combinations of metal streaks and striae also oriented in the same direction are present on the distal fragment (Fig. 7. 3 a). Viewed in this manner, the transverse break

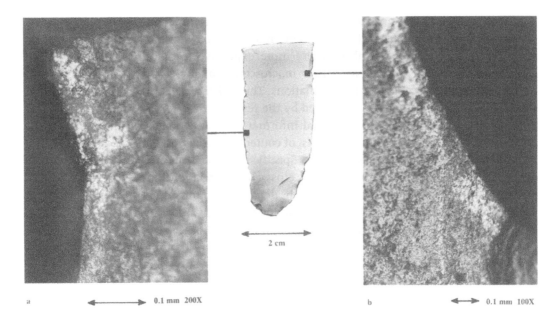

2 cm

Figure 7.2. Photomicrographs of metallic residues on blade 8.

would appear to have been caused by snapping of the blade as a metal machine bore down and came into contact with the ventral surface of the blade; the machine's direction of movement would have been parallel to the blade's longitudinal axis.

Another detail, although not as clearly associated with the breakage of the specimen, is the lateral edge damage. As shown in Figure 7.3 b, the shape of the flake facets is cuplike, and the facets are likely to have striae and abrasive particles parallel to their longitudinal axis. These would likely have occurred coincidentally with the edge damage; at the very least, they are consistent with the overall orientation of the conchoidal fracture. In this instance, it is interesting that the striae are contained within the flake facet and do not continue beyond its margins, which is not typical of tool use. The striae and abrasive particles would also be evidence of movement of the specimen perpendicular to the axis of mechanical failure and could have occurred as the result of minor movement within the soil profile before, during, or after specimen breakage. This artifact exhibits no wear traces that would be unambiguously associated with tool use; the artifact would appear not to have been used as a tool.

Blade 10, broken similarly by transverse snaps into four conjoinable pieces (Fig. 7. 4) from the subsurface, also exhibits wear traces indisputably caused by mechanical damage. Metal streaks and the now

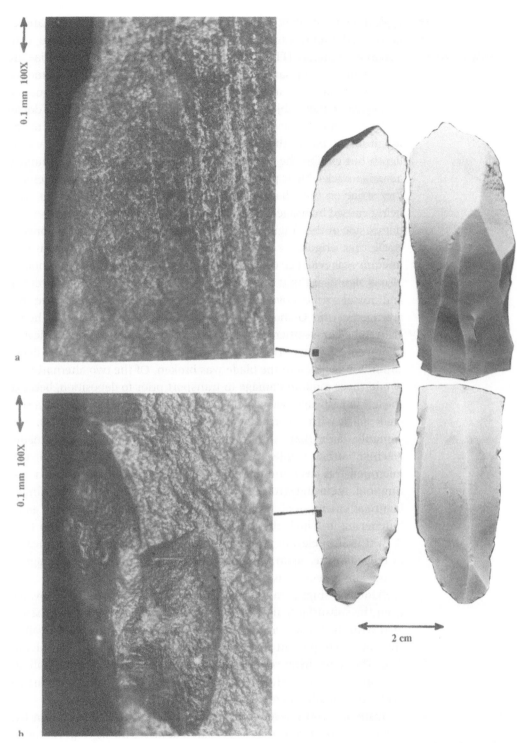

Figure 7.3. Blade 8 photomicrographs: (a) striae and metallic residues indicative of mechanical damage; and (b) edge damage.

typical broad, shallow, abrasive striae indicative of mechanical damage occur on the ventral surface at the points of breakage on adjacent, conjoinable fragments (Figs. 7.5 and 7.6 a). In this instance the striae are not oriented in the same direction on the two conjoinable fragments, indicating that as breakage occurred, the individual fragments rotated and were further subjected to direct mechanical abrasion. The dorsal surface of one of these fragments also displays striae parallel to the longitudinal axis of the blade (Fig. 7.6 b). These striae are of variable depth but exhibit the same general characteristics of the mechanical abrasive tracks. Their placement, orientation, and lack of complementary striae on the adjacent conjoinable fragments also rules out their being caused by usage of the blade as a tool. The opposite edge on the dorsal side of this fragment (Fig. 7.4 b) also has a mechanical abrasive track that originates at a cupped-out flake scar, which is further unequivocal evidence of edge damage caused by mechanical action. It also is significant that these mechanical abrasive tracks on the ventral and dorsal surfaces override individual, randomly oriented striae that are narrow and U-shaped in cross section and appear to be chemically etched or weathered. These are best interpreted as being indicative of movement either prior to burial or within the soil matrix after deposition but well before the blade was broken. Of the two alternatives, I would favor random damage in transport prior to deposition, because the left lateral edge (viewed dorsally) near the proximal end has a striated polish with unweathered striae (Fig. 7.4 a) that are, thus, presumably later than the randomly oriented and slightly broader "etched" striae. Unlike the other microwear already discussed, this micropolish is clear and unambiguous evidence of blade use as a cutting tool. A cutting stroke parallel to the blade edge is evident. Almost identical cutting-tool use wear is produced by experimental game butchering in which the tool edge comes in contact with bone (Fig. 7.7). It is also relevant to note that the blade 10 use wear is truncated by edge damage, again showing the sequencing of the edge damage as being subsequent to use.

Blade 5, a single, transversely broken proximal fragment excavated from the subsurface, also shows the same mechanical damage associated with the break (Fig. 7.8). Three discrete sets of mechanical abrasive tracks are present just proximal to the break on the ventral surface. The most distal set actually extends to the break on an oblique axis and converges with the two other sets. These tracks are further evidence of blade fragment rotation at the time of breakage.

Blade 3, a transversely broken distal fragment collected from the surface, has obliquely oriented, striated abrasive tracks (Fig. 7.9) adjacent to edge damage near the right edge, when viewed from the ventral surface. Distal to this spot are metal streaks that parallel the blade

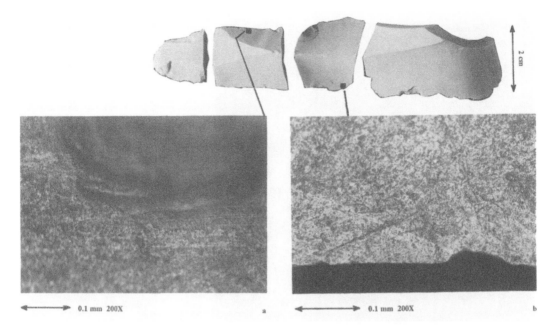

Figure 7.4. Photomicrograph of ventral surfaces of blade 10 fragments: (a) use-wear polish truncated by edge damage; and (b) mechanical-damage track overriding etched striae.

axis and are undoubtedly evidence of the cause of the transverse break. Similar metallic residues appear as streaks that originate at the edge damage on both sides of the ventral surface of blade 4, on both surfaces of blade 25 (Fig. 7.1 a), and on the dorsal surface of the unreattached flake (Fig. 7.1 b).

Blade 9 consists of two transversely broken conjoinable fragments (Fig. 7.10) excavated from the subsurface. Due to thickness and longitudinal axis curvature, the distal fragment was not examined microscopically. Collins had previously noted the proximal segment edge damage as likely being the product of machine plow damage, and the microwear analysis also supports his assessment. The characteristic abrasive tracks occur on both the dorsal and ventral surfaces of the proximal segment and are undeniable evidence of the cause of breakage. As shown in Figure 7.10, the photomicrographed area near the right edge as viewed from the ventral surface has a complex micropolish pattern of overlapping striae, or ones that show an abrasive sequence. The final abrasive track (which at 200X appears as two obliquely oriented divots beginning at the edge damage) in the lower right corner of the photomicrograph is another example of mechanical damage; it overrides the other striae, which intersect. Some of these parallel the blade axis, others are transverse or oblique to it. One

0.1 mm 200X

2 cm

Figure 7.5. Photomicrograph of ventral surface of blade 10 segment showing mechanical-damage track originating at the break and overriding etched striae.

way to interpret the earlier pattern of intersecting striae would be to attribute them to usage of the blade as a cutting tool. Were this true, then the typical cutting stroke would have entailed motions oblique to the edge as well as parallel to it, in what might be a back-and-forth rocking—or sawing—motion. This is perhaps the best interpretation, especially given that there is not a more general pattern of abrasion such as that produced over most flake arrises in experimental high-energy fluvial depositional environments (Kay 1997, n.d.).

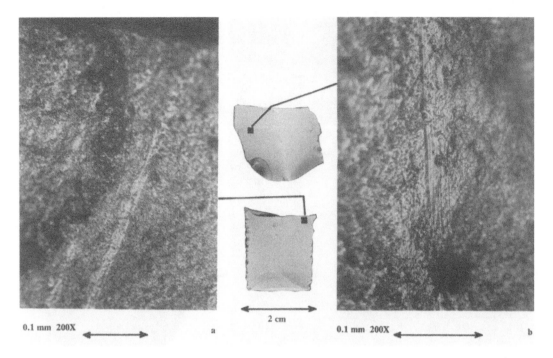

0.1 mm 200X   a

2 cm

0.1 mm 200X   b

Figure 7.6. Photomicrographs of (a) ventral and (b) dorsal mechanical-damage tracks on blade 10 conjoinable segments.

Blade 11, transversely broken into two conjoinable fragments and excavated from the subsurface, lacks the telltale mechanical abrasive tracks and conceivably could have been intentionally broken prior to caching. If so, it is the only cached specimen that cannot be categorically labeled as mechanically damaged. Blade 11 is also noteworthy for its tool-use wear (Fig. 7.11). Viewed from the ventral surface, the right edge of the proximal fragment has a consistent pattern of striated polish that forms a broad band at the edge only and extends from the proximal end to near the break. Striae are both parallel to the edge and—less commonly—transverse to it. Experimental controls show this kind of cutting-tool usage occurs on either a relatively hard substance or on something supported on a hard background. My preference would be cutting and scraping of hide laid out on either hard ground or supported by a flat rock or similar hard surface. Damage along the right edge would appear to be use-related and the likely source of the abrasive particles that striated the edge.

Blade 13, an obliquely fractured proximal fragment, was found on the surface, separate from the other specimens. It is distinct from them in being heavily patinated and also in its microwear (Fig. 7.12).

0.1 mm 400X

Figure 7.7. Experimental microwear produced in the butchering of a deer hindquarter.

A metal streak occurs proximal to the break on the ventral surface, and it is likely that the breakage of the blade also is due to mechanical damage. The dorsal surface microwear is one of striae in two directions, one parallel to the blade axis and the other oblique to it at the arris. The striae are not associated with the edge of the specimen. They are relatively old and have an etched appearance suggestive of long-term, post-depositional physical abrasion and chemical weathering.

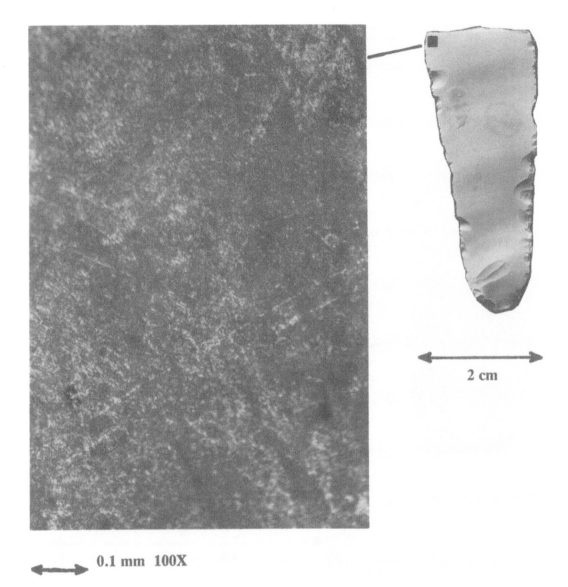

0.1 mm 100X

Figure 7.8. Photomicrograph of metallic residues on blade 5, showing three different directions due to rotation of specimen.

SUMMARY AND CONCLUSIONS

With the possible exception of blade 11, the most likely explanation for breakage of the specimens is recent mechanical tool damage. Mallouf's (1982) assessment of the Brookeen Creek blade cache as being plow-damaged would seem to be complemented by this analysis of the Keven Davis cache. Microscopic and macroscopic damage in the Keven Davis cache are expectably and demonstrably from metal machine damage—including much of the edge damage that Young

2 cm

0.1 mm 400X

Figure 7.9. Photomicrograph of striae probably indicative of mechanical damage on blade 3.

and Collins ascribed incorrectly to tool use. The excavated specimens from the Keven Davis cache occurred at variable depths. Blade 10 fragments, for instance, are noted as occurring from 6 cm to 39 cm below the surface spread over a horizontal area of 75 cm. The microwear evidence for mechanical-striated and metallic-residue abrasive tracks and the rotation within the soil profile of these fragments during breakage is consistent with damage caused by a metal plow, more so than by the final earth-moving equipment that ultimately uncovered

Figure 7.10. Photomicrograph of ventral surface of proximal fragment for blade 9, showing potential use-wear polish truncated by mechanical-damage track. Note also the plow damage on the opposite edge.

some of the items. Thus, the originally cached items were undoubtedly whole prismatic blades with relatively sharp, undamaged edges.

Of the total of fourteen individual blades recovered, ten of which were subjected to microwear analysis, utilitarian tool usage is unambiguous and clear only for blades 10 and 11, and is likely for blade 9. These use-wear results indicate the cached items were only incidentally used and were selectively stored for some later usage, either as cutting tools or for some other purpose. Blade 10 is also singularly

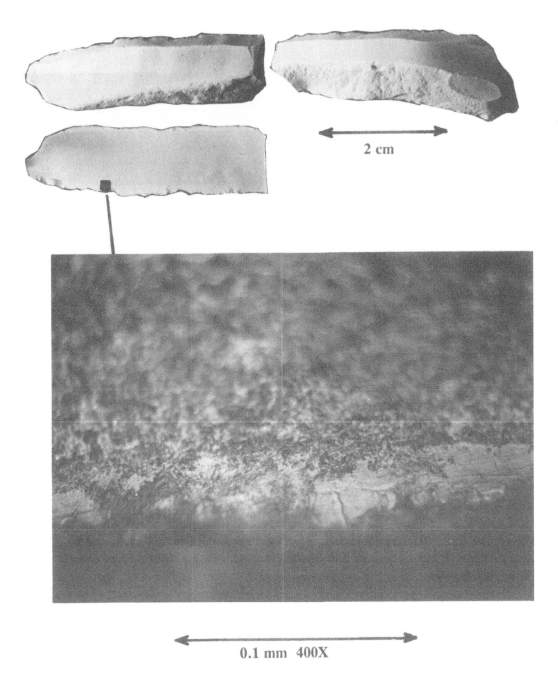

2 cm

0.1 mm 400X

Figure 7.11. Photomicrograph of use wear at edge of
ventral surface of blade 11 proximal fragment.

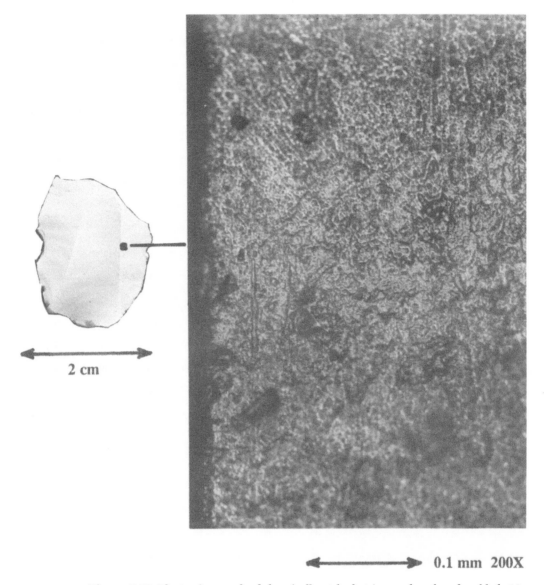

2 cm

0.1 mm   200X

Figure 7.12. Photomicrograph of chemically etched striae on dorsal surface blade 13.

noteworthy for having randomly oriented, weathered or chemically etched striae that may have preceded tool use and could be evidence of transport prior to both tool use and subsequent caching.

Blade 13, a surface find, appears unrelated to the cache. Its patination and wear traces suggestive of post-depositional chemical weathering are inconsistent with the other specimens, from which it was separated by approximately 150 meters.

PART THREE
Comparisons and Considerations of Caching Behavior

BLADES ARE KNOWN FROM NUMEROUS ARCHEO-
logical sites and cultures in the western hemisphere and,
more particularly, in North America, some clearly of early
Paleoindian affinity and others from numerous later cultural manifestations (Ford 1969). There is a great deal of variability within and
between blade assemblages, and there is no established delineation or
constrained range of variation for "Clovis blades." These conditions
make it difficult to identify confidently the cultural affinity of specimens from uncertain contexts, such as the isolated Keven Davis cache.

Any effort to extract comparative data on prismatic blades from the
archeological literature of fluted-point sites in North America faces a
number of uncertainties, ambiguities, and misnomers. Uncertainty
arises, for example, when it is unclear whether a site report that does
not mention blades is accurate or somehow fails to cover blades that
were, in fact, part of the assemblage. Ambiguity is found, for example,
in reports that discuss blades but the illustrated examples do not
appear to be true blades or, conversely, in reports in which blades are
illustrated but not discussed. Terms such as "flake knives," "lamellar
flakes," or "Old World blades" have unclear meanings, but the most
common kind of misnomer is use of the word *blade* to refer to bifacially flaked objects. It is instructive to review some specific cases to
underscore my expectation that with better reporting the distribution
of Paleoindian blades will become much clearer.

The reports of numerous early sites lack any mention of blades or
tools made on blades, and nothing in the discussions or illustrations
indicates any evidence to the contrary. Examples of such reports are
those for Vail in Maine (Gramly 1982; Gramly and Rutledge 1981);
Bull Brook II in Massachusetts (Grimes et al. 1984); Leavitt in
Michigan (Shott 1993); Kimmswick (Graham et al. 1981) and Rodgers
(Ahler and McMillan 1976) in Missouri; Whipple in New Hampshire

(Curran 1984); Potts (Gramly and Lothrop 1984) and Shawnee Minisink (McNett 1985; McNett, McMillan, and Marshall 1977) in New York; Debert in Nova Scotia (McDonald 1966; Byers 1966); and Alder Creek (Timmins 1994), Crowfield (Deller and Ellis 1984), Cummins (Julig 1984), Fisher (Storck 1983), and Parkhill (Roosa 1977) in Ontario. In a few cases, such as Rodgers Shelter in Missouri, it is possible that blades are not reported because the analyses were incomplete at the time of the writing, but for most of these, it is to be assumed that blades are not a part of the assemblages. In contrast to the ambiguous reports, one can be particularly confident in the Debert site report by McDonald (1968) because he specifically notes the absence of blades and attributes that absence to limitations in raw materials.

In addition to site-specific reports, some regional syntheses do not refer to blades or tools made on blades, as for example those for the lower Great Lakes region (Ellis and Deller 1988) or the Hudson Valley (Funk 1977), whereas other regional treatments do, such as Purdy's (1981:37–38, 42, and Plate 4) survey of prehistoric stone technology in Florida.

For some sites, differing accounts appear in different sources or there is contradictory information within a single report. The Williamson site in Virginia was first described by McCary (1951) without any mention of blades or blade cores but is later noted by Cox (1986) as a prime example of evidence for an early blade-making technology; Cox speaks of there being numerous blades as well as cores in the collections from Williamson. Also, at Graham Cave, Missouri, blades from the earliest levels are illustrated by Logan (1952), who referred to them as "flake knives," but are not mentioned or illustrated in an update on the site by Klippel (1971).

Among sites with convincingly reported blades or tools made on blades are Stanfield-Worley in Alabama (DeJarnette, Kurjack, and Cambron 1962), Michaud in Maine (Spiess and Wilson 1987), Bull Brook in Massachusetts (Byers 1954), Gainey (Simmons, Shott, and Wright 1984) and Holcombe Beach (Fitting, DeVisscher, and Wahla 1966) in Michigan, Plenge in New Jersey (Kraft 1973), Corditaipe in New York (Funk and Wellman 1984), Thedford II in Ontario (Deller and Ellis 1992), Wells Creek in Tennessee (Dragoo 1973), and Reagen in Vermont (Ritchie 1953). However, none of these, except perhaps those from Stanfield-Worley and Wells Creek, is the product of a robust blade technology. Instead, these are pieces that barely meet the definition of blade, tend to be small, and often occur in limited numbers. The Shoop site in Pennsylvania is an instructive example of the marginal nature of some of these "blade" assemblages or specimens. Whitthoft (1952) based his Enterline chert industry in large part upon the evidence for blade manufacture at the Shoop site, but his artifact illustrations (Whitthoft 1952: Plates 2 and 3), as well as his schematic illustration of blade reduction (Whitthoft 1952: Fig. 2), do not suggest a well-developed blade technology (although several of the illustrated artifacts do appear to be made on blades). Byers (1966: 37–38) and Ritchie (1965: 30) expressed doubts regarding the evidence for Enterline as a core-and-blade industry. Cox (1986: 137) states

> *The people at Shoop did know how to consistently make blades, and judging from the evidence at Williamson, probably did have prepared blade cores. However, as far as blade-making is concerned, Shoop is certainly a rather pale reflection of the upper Paleolithic blade industries, and the bulk of its tools was probably made from flakes struck off relatively unspecialized flake cores and large blanks.*

"Blades" or tools made on "blades" are reported from Bostrom in Illinois (Tankersley 1995), Paw Paw Cove in Maryland (Lowery 1989), and Barnes in Michigan (where one specimen is described as a "morphological blade" [Wright and Roosa 1966: 857]), but the accompanying illustrations in each of these cases suggest otherwise. Of course, an illustration is not as good as hands-on inspection, and the authors' classifications may be correct.

Adovasio (1993: 208–211) reports "small, prismatic blades" from the stratum IIa Paleoindian component at Meadowcroft Rockshelter and from the nearby Krajacic site, where polyhedral blade cores also were found. These are indeed small (in fact, they were earlier reported as "microblades" [Adovasio et al. 1977: 152]), but their size range exceeds most standard microblade definitions. These blades from western Pennsylvania are smaller and straighter and have larger bulbs and proportionally larger platforms than Clovis blades. They also predate Clovis and, therefore, are among the important clues to be considered in tracing the history of blade technology in the Americas.

It should be clear from the foregoing that mainstream archeology in North America is ambivalent toward blade technology. There seems to be a sense that blade making could be important, particularly to Paleoindian studies, since it may have remote ties to Old World Upper Paleolithic blade technology, but it has received very inconsistent treatment. Blade technology can be described as everything from thoroughly analyzed and reported to erroneously identified to overlooked in the literature.

Selected data on prismatic blades from known Clovis or possible Clovis contexts, as well as from definitely non-Clovis affiliation, are here assembled as a comparative base for assessing on morphometric grounds the affinity of the Keven Davis blades to blades of Clovis affiliation. Nine blades from the Keven Davis cache are sufficiently complete for these comparisons. Data on blades from twenty-four additional localities have been assembled (Table 6.1) mostly from the literature, but also from unpublished collections in a few cases. The data available for these collections vary significantly, and all measurements and observations cannot be tabulated here for all data sets. In the case of four sites, only averages of measurements are reported. For three sites, it was necessary to estimate blade measurements from illustrations, and these estimates, as reported in Table 6.1, are therefore to be taken as approximate. This comparative study does not address microblades, defined by de Heinzelin de Braucourt (1962: 438) as being less than 3 cm in length.

Observed attributes as well as measurements and indices, as available or determinable, are the basis of these comparisons. Observational information is discussed in the text and metric comparisons are displayed in graphic form. Metrically, the nine complete Keven Davis blades are first compared with the other data sets through two-dimensional scatter graphs of length and width, which depict the absolute sizes and proportions of blades (Figs. 8.1–8.5). Next, the relative proportions of the three principal dimensions are displayed on triangular coordinate graphs, a visual way to compare shapes (Figs. 8.6–8.10).

BLADES FROM CLOVIS CONTEXTS
BLACKWATER DRAW

Blackwater Draw produced the first Clovis blade assemblage to be recognized and described as such (Green 1963), and for over thirty years the data on those blades as detailed by Green have been the nearest thing to a type description or definition. Green described seventeen specimens and inferred from adhering matrix and other circumstances of the discovery that their original provenience had been in the Clovis-age Gray Sand. He also inferred that the blades had been cached. Most of the blades recovered by Green had been broken by machinery (Green 1963: 150). Mallouf (1982: 96–97) also observes that much of the edge modification and breakage on these specimens is comparable to that on the Brookeen specimens and was probably caused by mechanical equipment operating in the gravel quarry.

Other blades of secure or probable Clovis age are known from this site (J. Hester 1972: 97 and Figs. 88, 89, and 91–95). Several of the specimens reported by Hester are blade segments. One unusually large side scraper apparently made on a blade (J. Hester 1972: Figs. 93 f and 94 i) is 165 mm long.

Five specimens found more recently (Montgomery and Dickenson 1992: 32–33) are similar but include one exceptionally long blade (215 mm; artifacts 1 and 2 conjoined).

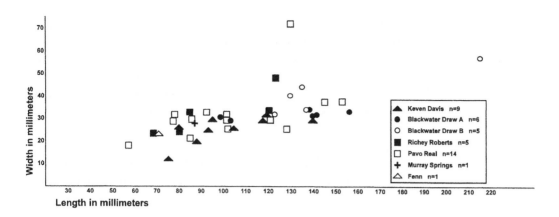

Figure 8.1. Scattergram of blade lengths and widths, comparing Keven Davis blades with those from Clovis contexts.

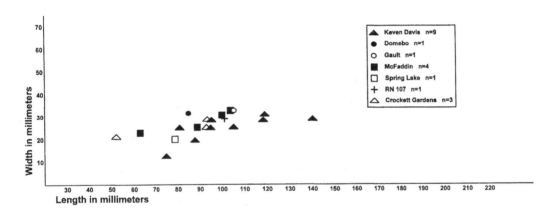

Figure 8.2. Scattergram of blade lengths and widths, comparing Keven Davis blades with those from additional Clovis contexts.

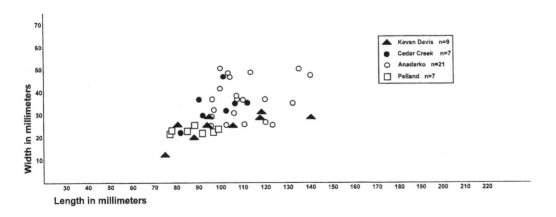

Figure 8.3. Scattergram of blade lengths and widths, comparing Keven Davis blades with those from indefinite contexts.

The seventeen specimens found by Green are nearly identical in all respects to those from the Keven Davis cache in terms of platforms, blade exteriors and interiors, and shapes. Six of the original seventeen blades are sufficiently complete for metric comparisons. Only in the case of size (Fig. 8.1) is there a distinction, with three of the Blackwater Draw specimens being longer than any of the Keven Davis pieces; also, five of the Keven Davis pieces are shorter than any of the complete specimens reported by Green. Shapes (Fig. 8.6) are virtually identical except for Keven Davis blade 4, which is shorter, wider, and thicker than average; it is also a secondary cortical blade of removal Stage 2.

Two of the five blades reported by Montgomery and Dickenson compare closely with both the Keven Davis and the Blackwater Draw specimens reported by Green, two are similar in length but are wider, and one is significantly longer (Fig. 8.1). Two groups of shapes emerge when these five blades are plotted on the triangular graph (Fig. 8.6), with two specimens being shorter, wider, and thinner than eight of the Keven Davis blades and three specimens being thinner and wider.

EAST WENATCHEE

Five blades from the East Wenatchee Clovis cache in central Washington are illustrated by Gramly (1993: 45, 50). No descriptions or metric data are provided, but it is possible to discern attributes from the excellent drawings by Valerie Waldorf, and even to estimate measurements (Table 6.1). These blades have tiny platforms, smooth interiors, almost no bulbs of percussion, strong curvature, and triangular as well as trapezium-like cross sections. Since a ditching machine hitting a portion of the cache led to its discovery (Gramly 1993:5), it is not entirely clear if the edge modification illustrated on the blades is completely prehistoric retouch or if any metal tool or machine damage is present. The preliminary information available on this site does not allow definitive conclusions, but most, if not all, of this edge modification seems to be prehistoric (Gramly 1993: 7).

Metrically, the five East Wenatchee cache blades, in comparison with the Keven Davis blades, are similar in their range of absolute lengths, but tend to be wider (Fig. 8.1). This tendency is more apparent when thickness is also considered (Fig. 8.6). The East Wenatchee blades are proportionally shorter than all but blade 4 from the Keven Davis cache.

PAVO REAL

A sealed stratum containing early Paleoindian artifacts was documented at the Pavo Real site in central Texas (Henderson and Goode 1991: 26–28). Folsom points and preforms as well as Clovis points were found along with considerable flaking debris and other artifacts. Prominent among the materials from Pavo Real are prismatic blades, blade cores, scrapers and burins on blades, and debitage from blade production. It is not clear how the Clovis and Folsom diagnostic artifacts came to rest in the same deposit at Pavo Real, but the blade technology is so similar to that seen in other Clovis components and, to my knowledge, never seen in Folsom components, it is here assumed that the blades, blade cores, and blade related debitage are of Clovis affinity. The site is adjacent to an outcrop of stream cobbles and boulders of good-quality Edwards chert, and the entire blade-core reduction sequence is reflected in the cores, blades, and debitage. Blades in removal stages 1 through 6 were recovered. On a very few pieces there is edge damage clearly identifiable as occurring during excavation;

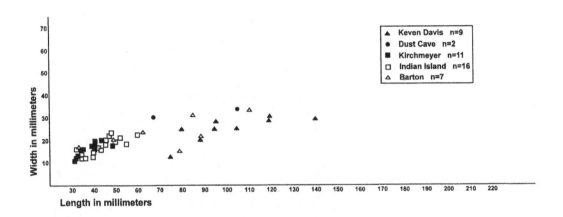

Figure 8.4. Scattergram of blade lengths and widths, comparing Keven Davis blades with those from non-Clovis contexts.

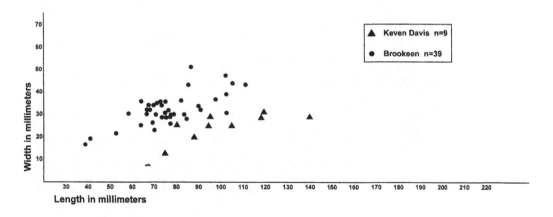

Figure 8.5. Scattergram comparing lengths and widths of blades from the Keven Davis and Brookeen caches.

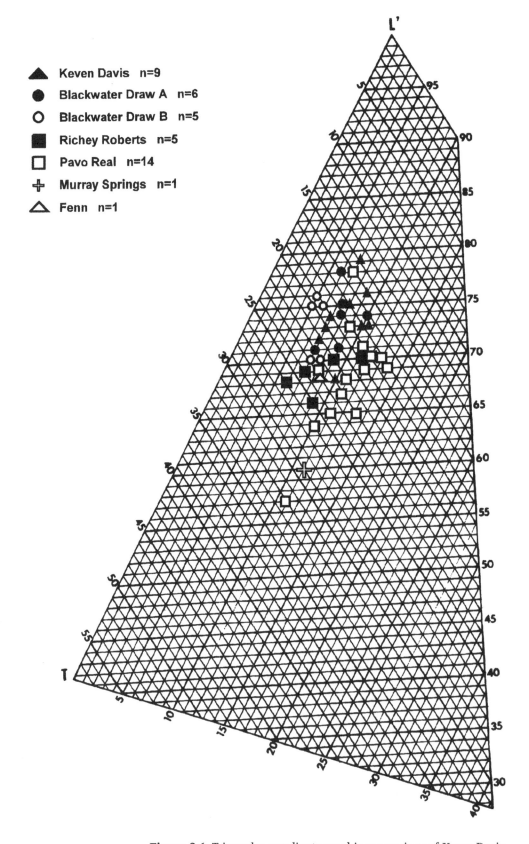

Figure 8.6. Triangular coordinate graphic comparison of Keven Davis blades with those from Clovis contexts.

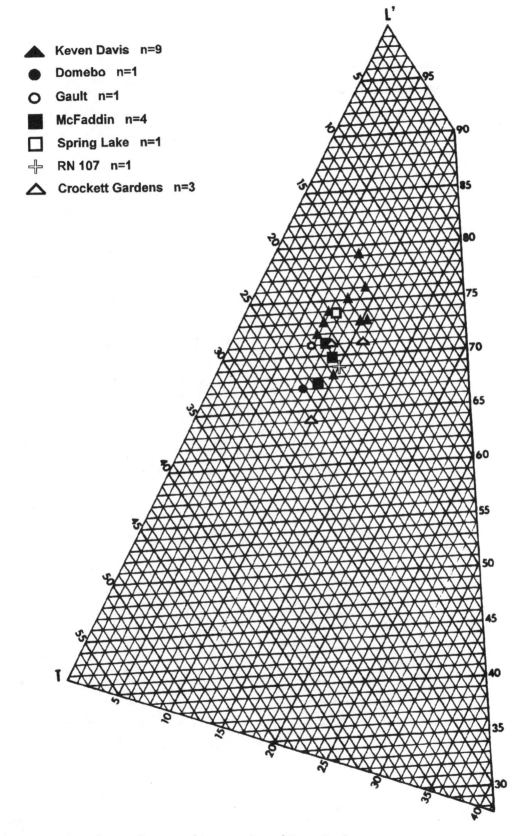

Figure 8.7. Triangular coordinate graphic comparison of Keven Davis blades with those from additional Clovis contexts.

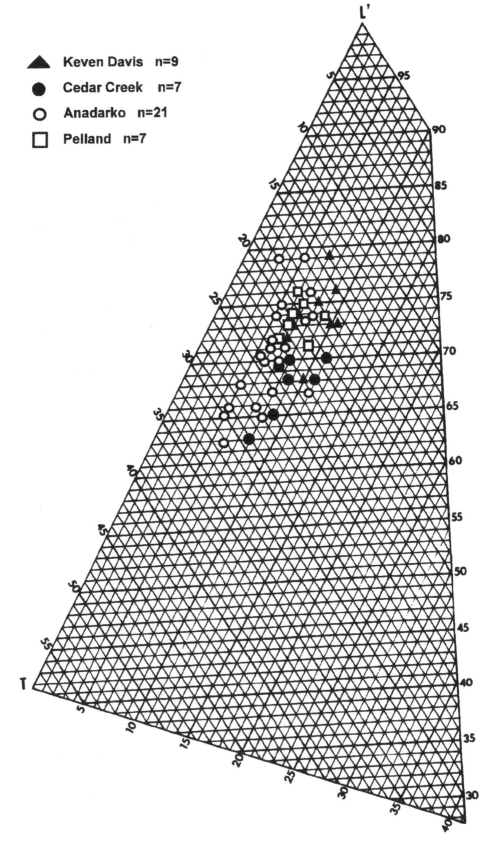

Figure 8.8. Triangular coordinate graphic comparison of Keven Davis blades with those from indefinite contexts.

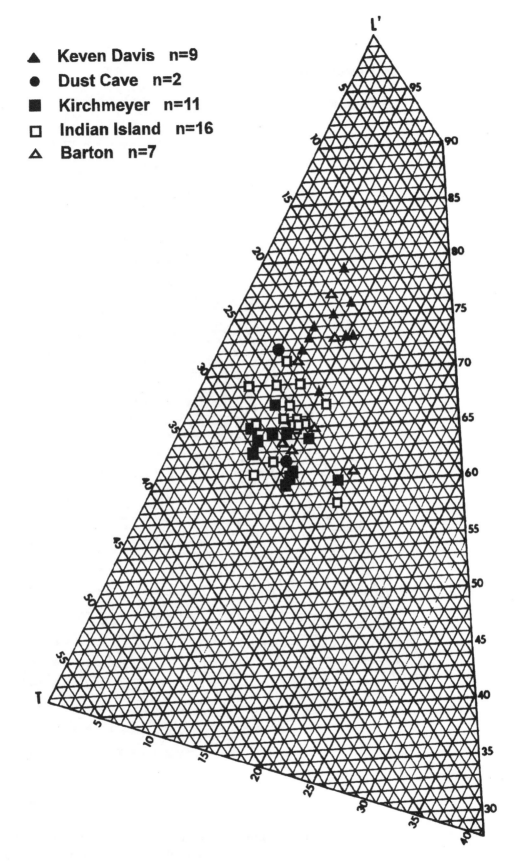

Figure 8.9. Triangular coordinate graphic comparison of Keven Davis
blades with those from non-Clovis contexts.

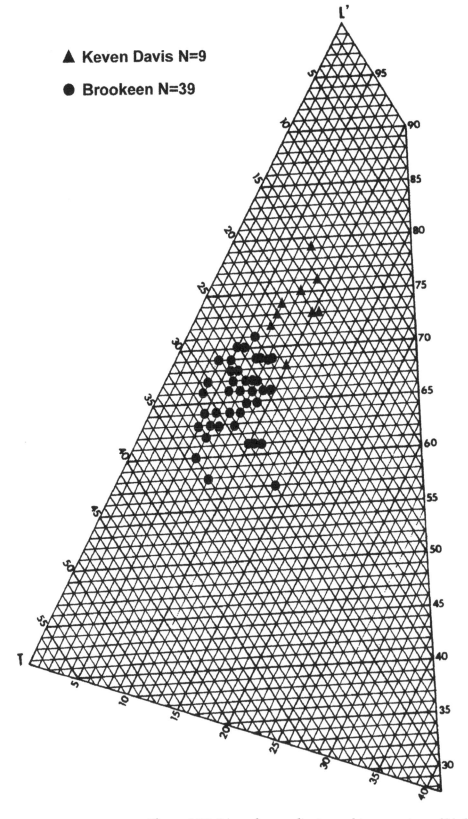

Figure 8.10. Triangular coordinate graphic comparison of blades from the Keven Davis and Brookeen caches.

otherwise, the breakage and edge modification are prehistoric. Glenn
Goode in a preliminary analysis of the lithic material from Pavo Real
isolated thirty-three blades and blade segments. The segments were
broken prehistorically, either accidentally or purposefully.

Fourteen blades from Pavo Real yield complete metric data. Not surprisingly, given the range of reduction stages represented, there is considerable variation in sizes and shapes among these specimens.
Absolute length and width data, with the exception of that for one
particularly wide blade, scatter with and relatively near the Keven
Davis blades (Fig. 8.1). Blade shapes, too, are well within the range
of variation seen in the Keven Davis pieces, but the Pavo Real blades
show a wider dispersion (Fig. 8.6), especially toward shorter and
thicker proportions. Contrary to what might be expected, these shorter
and thicker blades are not consistently from early removal stages.

Since Pavo Real is the only assemblage examined in this study that
represents a blade-production workshop not ravaged by plowing (as
was Adams, the only other blade workshop), the relatively high variability among Pavo Real blade attributes and the high proportion of
prehistoric breakage (19 out of 33, or 58 percent) are probably repre-
sentative of the full ranges of variability and breakage that occur in
blade production. The other assemblages, in contrast, likely reflect
selectivity of blades for certain specifications and completeness.
Where machinery—or the trampling of large animals—has not caused
breakage, complete blades prevail.

At the Murray Springs site in southern Arizona (Hemmings 1970; ## MURRAY SPRINGS
Huckell n.d.) thirteen blades and tools made on blades along with
Clovis points were found in association with remains of mammoth
and other animals. Huckell (n.d.: 87–97) notes strong similarities
between the Murray Springs blades and those described by Green
from Blackwater Draw. He also reports that two of the blade tools at
Murray Springs are large (termed "heavy blades") with marginal
retouch and notes that two specimens from the nearby Lehner Clovis
site (Haury, Sayles, and Wasley 1959: Fig. 14 a and 14 b) are possibly
fragmentary tools made on "heavy blades." The other Murray Springs
blades as described and illustrated by Huckell have the small platforms, flat bulbs, and low-amplitude interior ripple marks seen in
other Clovis blades. A badly damaged core and some blade-core
debris suggest that some blade detachment occurred at the Murray
Springs site.

One blade specimen from Murray Springs is sufficiently complete to
provide comparative metric data (Table 6.1). That blade is close to the
lower end of the range of variation in length and width seen in the
Keven Davis blades (Fig. 8.1). In overall proportions, the Murray

Springs blade is relatively close to Keven Davis blade 4—that is, short—but it is also more typical of the Keven Davis cluster in terms of thinness and width (Fig. 8.6).

HORN SHELTER 2

In the lowest unequivocal cultural layer (Stratum 3) of Horn Shelter 2 in central Texas were found a blade and a blade segment along with a small amount of other cultural debris, hearths, bones of turtle and other animals, but no Clovis points (Redder 1985; Story 1990, Watt 1978). There is a single radiocarbon date of 10,150 ± 120 B.P. (Tx 2189; Watt 1978: 131–132) on turtle shell apatite—and, therefore, a minimal age—from Stratum 3. An overlying stratum (5a) yielded a Folsom point fragment and a complete point resembling Suwanee points from Florida. The latter are thought to have an age similar to that of Clovis. The blade from Stratum 3 (Fig. 8.11 a) is made of Edwards Chert and has the typical Clovis attributes of smooth interior, small bulb, and small platform. It is triangular in cross section and the exterior ridge is battered. The distal end of the blade is missing as the result of a break evidently caused by a flaw in the chert. There is minor, prehistoric nicking of the blade edges. Metrically (Table 6.1), the Horn Shelter blade is 19 mm wide and 7.3 mm thick; it was originally greater than 75 mm long.

FENN CACHE

A single blade is reported from the Fenn cache (Frison 1991: 331, Fig. 19.12). This major assembly of Clovis artifacts is of imprecisely known origin—probably originating from somewhere near the southwestern corner of Wyoming or northeastern Utah—and is inferred to have been a cache primarily on the basis of similarities with the known Clovis caches Anzick, East Wenatchee, and Simon. The Fenn cache blade has a small platform, slight bulb, strong curvature, and multiple prior blade scars on its exterior, and it conforms to other Clovis blades in all respects. Metric data on the Fenn cache blade presented here (Table 6.1; Figs. 8.1 and 8.6) are estimated from the scale drawing provided by Frison. The blade is shorter and wider than the Keven Davis blades (Fig. 8.1); it is virtually identical to the single measurable blade from Murray Springs and one of the blades from East Wenatchee in its overall proportions (Fig. 8.6), which compare favorably to blade 4 from Keven Davis.

ADAMS

The Adams site in western Kentucky yielded an impressive surface collection of deeply patinated lithic artifacts with strong technological and typological affinities to Clovis (Sanders 1990). There is a significant number of blade cores and blades from the site and evidence that, as in sites in the south-central United States, large bladelike flakes were among the forms reduced bifacially into Clovis points (Sanders

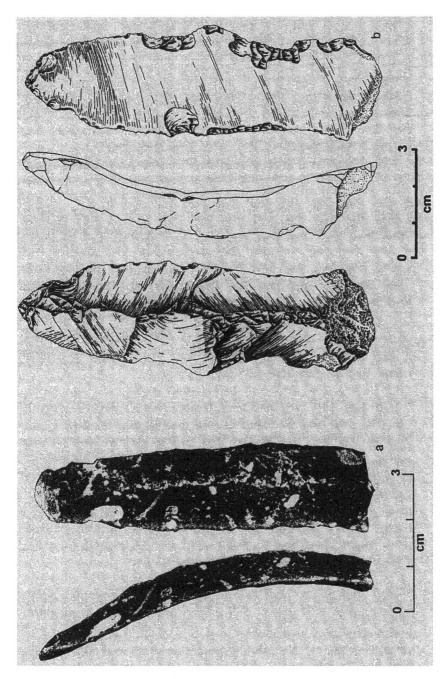

Figure 8.11. Blades with Clovis attributes from Horn Shelter and site 41RN107.

1990: Figs. 5, 11, and 12). Sanders (1990: 60, Tables 1 and 2, and Figs. 41–47 and 54) reports forty unmodified blades, provides summary metric data on the seven complete specimens, and illustrates several blades, blade fragments, and modified blades and blade segments. From his illustrations of the Adams site blades, it would appear that some of the modifications that Sanders identifies as intentional and classifies into various tool forms are actually plow damage as diagnosed by Mallouf (1981, 1982) for the Brookeen blades.

Technological attributes of the Adams blades show a range of variation from early, cortical to late, noncortical removals. The illustrated Adams specimens tend to have larger platforms and more pronounced bulbs than do the Keven Davis and Blackwater Draw specimens. Other attributes, such as blade curvature, are not discernible from the report.

Since metric data on individual blades are not provided, the averages reported by Sanders (1990: Table 2) can be used only for generalized comparison. In terms of average lengths and widths, the Adams blades are toward the shorter and wider end of the scatter of known Clovis blades, comparing most closely to the blades from Domebo and Murray Springs. In terms of proportions, the Adams blades fall more in the range of shapes of Late Prehistoric blades (from the Indian Island, Barton, and Brookeen sites in Texas).

BLADES FROM PROBABLE CLOVIS CONTEXTS
DOMEBO

Hammatt (1969) reports a blade found in the stream bed below the spot where the Domebo mammoth and associated artifacts were excavated in Oklahoma (Leonhardy 1966). The piece is comparable in most attributes to those of definite Clovis affiliation (prior blade scars on its exterior, smooth interior, flat bulb, and small platform, but little curvature) and has distinct retouch along one lateral edge. The metric data on this blade show it to be well within the length range of Clovis blades, but it is toward the wider sector of the scatter (Figs. 8.2 and 8.7), where it compares closely with one blade each from East Wenatchee, Murray Springs, and Pavo Real.

GAULT

The Gault site in central Texas (Collins et al. 1991; Collins, Hester, and Headrick 1992; Hester, Collins, and Headrick 1992) yielded three blade segments in association with a Clovis point (Collins, Hester, and Headrick 1992: 4). The attributes visible on these segments (e.g., Fig. 3.18 b) are like those seen on Blackwater Draw and Keven Davis pieces in terms of very small platform, smooth interior, strong curvature, and multiple prior blade scars on the exterior. There is very little question that these are Clovis blades; however, none is complete and they yield no meaningful metric data. An overlying stratum yielded a large polyhedral blade core along with mixed Paleoindian and early Archaic artifacts; it, too, is inferred to be of Clovis affiliation.

Other possible Clovis blades, blade cores, and blade-core debris are known from disturbed contexts at this site, including one complete blade excavated in 1929 by J. E. Pearce (collections and records housed at the Texas Archeological Research Laboratory). This specimen is patinated and has minor edge modifications along both margins, apparently of prehistoric origin. It shares the attributes of small platform, smooth interior, curvature, and multiple prior blade scars with known Clovis blades, although the platform is not as minute as some. Metrically, the blade collected by Pearce is comparable in absolute length and width (Fig. 8.2) to blades from Keven Davis, Blackwater Draw, McFaddin Beach, and Pavo Real. In overall proportions, it compares favorably with blades from Keven Davis and two of the recently collected blades from Blackwater Draw (Fig. 8.7).

McFADDIN BEACH

McFaddin Beach on the upper Gulf Coast of Texas has been the scene of numerous surface finds of fossil animal bones, Clovis points, Clovis blades, and other artifacts brought ashore by wave action (Hester et al. 1992; Long 1977; Turner and Tanner 1994; notes on the collections by Paul Tanner, on file, Texas Archeological Research Laboratory). At least eleven blades, blade segments, and retouched blades have been documented by Tanner; some of these are also discussed and illustrated by Long (1977: 10 and Figs. 3, 6, and 7). Data on three of these, as recorded by Tanner, are provided here (Table 6.1). Because of the high number of Clovis points and the similarity between the McFaddin Beach blades and those of known Clovis origin, it seems probable that these are Clovis in age. They also share attributes in common with known Clovis blades. Metrically, two of the three are similar to Blackwater Draw, Pavo Real, Gault, and Keven Davis blades, and one is near the short-wide extreme of variation seen in Clovis blades (Fig. 8.2). In terms of overall proportions, two fall within the spread of Keven Davis blades and one is shorter, wider, and thinner (Fig. 8.7).

SPRING LAKE

Spring Lake, a submerged site in central Texas, has yielded a large assemblage of prehistoric artifacts, including many of Paleoindian type (Johnson and Holliday 1984; Shiner 1981, 1982, 1983; Takac 1991), though as yet none of the Paleoindian material has been recovered from coherent deposits (Takac 1991: 46). Among the diagnostic points are at least three that are typed as Clovis (Takac 1991: 46), and a single prismatic blade has been noted in the collection (Table 6.1). The blade is within the length-width range of the shorter Keven Davis blades (Fig. 8.2) and is completely within the spread of shapes seen in Keven Davis blades (Fig. 8.7).

SITE 41RN107   This unnamed locality is an open site in the Colorado River valley of west-central Texas. Early Archaic, Folsom, Midland, and Plainview artifacts, along with the tip of a retouched and deeply patinated blade, have been collected from the surface of the site (Bryan and Collins 1988). Another deeply patinated blade found on the surface, having recently eroded from the fill of a spring-fed pond, has attributes like those of known Clovis blades (Fig. 8.11 b). The platform is small and multifaceted and the interior surface is quite smooth. The bulb is larger than that on most Clovis blades. The edges are nicked and chipped extensively, possibly by cattle or bison trampling in the moist pond fill. Metrically, this blade compares closely with the one complete blade from the Gault site (Fig. 8.2); its proportions are close to the shortest blade from the Keven Davis cache (Fig. 8.7).

CROCKETT GARDENS   The Crockett Gardens site (41WM419) was discovered during mechanical excavation of a large borrow pit adjacent to the North San Gabriel River in central Texas. Very little of the site remained at the time of archeological investigation, and extremely limited inquiry of a hasty nature was all that could be accomplished. These minor investigations revealed remnants of late Paleoindian, as well as early through late Archaic, components in addition to a deeply buried deposit (Stratum 6 in backhoe trench 5) from which were collected three prismatic blades clearly of Clovis-like morphology (McCormick 1982: 12.135–12.166). The geologic context of these three blades appears to be analogous to that of the Clovis component at the nearby Wilson-Leonard site, which is located along a tributary of the San Gabriel River and occupies a geomorphological setting nearly identical to that of the Crockett Gardens site. At both of these sites, the deepest cultural materials were found on top of the basal gravels in thick valley fills and were overlain by late-Paleoindian through Archaic sequences.

Surface collecting from disturbed contexts in and around the Crockett Gardens borrow pit recovered a fragmentary Clovis point, two additional blades and seven unfluted, lanceolate dart points (called "Plainviews" by McCormick), one of which is almost certainly a Midland point (McCormick 1982: Fig. 12.21 7 a). No detailed descriptions or metric data are provided for the blades, but line drawings to scale (McCormick 1982: Fig. 12.21 3 a–e) allow some general observations and metric estimates to be made. Of the three blades from the backhoe trench (McCormick 1982: Fig. 12.21 3 a, c, e), two are complete and one is broken. The two complete specimens are moderately curved and have edge modification (prehistoric retouch?) along the margins near the tips. One has subparallel exterior flake scars, whereas the arris on the other is formed by transverse (core-preparation?) scars as well as subparallel, longitudinal scars. The blade fragment from the backhoe trench is triangular in cross section and has two parallel exterior scars; it exceeded 102 mm in length. One of the two blades collected from disturbed contexts is small and has a triangular cross section; the other is a blade segment, prismatic in cross section with subparallel exterior scars. It exceeded 98 mm in length. All five of the Crockett Gardens blades lack cortex. The four blades with proximal ends present have small, nearly flat bulbs and appear in the drawings to have very small platforms.

Metrically these blades compare with the shorter, wider specimens from Pavo Real and Keven Davis, as well as with blades from the western Clovis sites (Fig. 8.2). Shapes of two of the complete Crockett Gardens blades are within the range seen for blades from Blackwater Draw and Keven Davis, while that of the other one is shorter and wider than most Clovis blades (Fig. 8.7).

Miscellaneous finds of probable Clovis blades worth mentioning include two as yet unpublished sites. Chris Ringstaff (personal communication 1992) documented a patinated, complete prismatic blade from a surface context in Llano County, west-central Texas. In every respect, this blade conforms to the attributes of blades from secure Clovis contexts. Prewitt (personal communication 1995) describes two blades that he considers to be of Clovis affinity from a surface site in Tyler County, extreme southeastern Texas. These are both of Edwards chert, both are strongly curved longitudinally, and the characteristic smoothness of the interior surfaces is seen on both. Dimensions supplied by Prewitt (personal communication 1995) are as follows: specimen 1: length 135 mm, width 29 mm, thickness 15 mm, width-to-length ratio 1:4.65, and platform angle 128 degrees; specimen 2: length 107 mm, width, 37 mm, thickness 9 mm, width-to-length ratio 1:2.89, and platform angle 129 degrees.

BLADES FROM INDEFINITE CONTEXTS
CEDAR CREEK

Hammatt (1969) reports seven prismatic blades from a locality on Cedar Creek in Washita County, western Oklahoma. These he attributes to "Paleoindian" origins on the basis of morphology, in the absence of secure contextual evidence. When compared to blades of known Clovis affinity, these seven are less regular, have more random exterior flake patterns, are straighter (except for one), and are in the shorter end of the Clovis range of lengths, and many of them tend to be wider than most Clovis blades (Fig. 8.3). On the other hand, they have the flat bulbs and smooth interior surfaces so characteristic of Clovis blades. These may or may not be Clovis blades, but if they are, they would be stage 4 or 5 removals. Proportionally (Fig. 8.8) these blades tend to overlap with the relatively shorter Keven Davis specimen (blade 4) but overall are shorter.

ANADARKO (McKEE)   Another assemblage of blades from Oklahoma is reported by Hammatt (1970) from near Anadarko. These blades have the small bulbs, very small platforms, smooth interiors, and pronounced longitudinal curvature seen in Clovis blade assemblages. In the illustrations, the edges of most of the Anadarko cache blades appear to lack significant nicking or retouch; however, Hammatt (1970: 145) reports that the "great majority . . . have signs of wear along the edges, and in many cases there has been a deliberate attempt to dull one edge of the blade, either by removing a notch or by applying secondary retouch." The Anadarko blades were excavated from a corral where erosion had exposed the top of the cache at the surface (indicating that they had been near the surface and possibly damaged by cattle trampling). At least eleven (50 percent) are broken, but it is unclear whether this is the result of damage since they were cached or if they were cached in pieces.

Twenty-one blades from the Anadarko cache yielded complete metric data. These blades are comparable in length to the Keven Davis assemblage, but in width some are similar to, and most are wider than, the Keven Davis blades (Fig. 8.3). This emerges clearly in the proportional depiction (Fig. 8.8). As a group, the Anadarko blades are thinner, wider, and shorter than the Keven Davis group.

PELLAND   Stoltman (1971: 105–109) reports eight prismatic blades (the "Pelland blades") from disturbed context in northern Minnesota. The blades were found under circumstances which might indicate that they were once cached along with an end scraper and at least three additional blades (one being a fragment). Associational data are meager, but the locality appears to be preceramic, and some lanceolate Paleoindian points have been found in the vicinity. More indicative of the possible Clovis affinity of the Pelland blades than contextual evidence is their close technological similarity to the blades reported by Earl Green from Blackwater Draw (Stoltman 1971:108). As Stoltman notes, the blades are quite similar except that the Pelland blades are shorter. Data on other Clovis blades not available at the time of Stoltman's report show even closer morphological similarity to the Pelland blades. The Pelland blades compare closely, for example, with Clovis blades from the Adams, East Wenatchee, Fenn, and Murray Springs sites in size, shape, lack of prominent bulbs, low-amplitude ripple marks, and small platforms. When compared metrically, the Pelland specimens compare favorably with the shorter Keven Davis blades in terms of length and width (Fig. 8.3). Shape as plotted on the triangular graph (Fig. 8.8) compares closely among the Pelland and Keven Davis assemblages. A significant difference in these comparisons is extensive pressure flaking on all of the Pelland blades, and it is not known how much this flaking may have reduced the width of the original blades.

Blades and tools on blades or bladelike flakes are often noted from Paleoindian sites in the Middle Tennessee Valley, although most have been recovered from surface contexts (Adair 1976; Broster and Norton 1993; Hubbert, n.d., 1989; Charles Hubbert, personal communication). Recent excavations at Dust Cave in northwestern Alabama (Driskell 1994; Meeks 1994) have brought a small assemblage of such artifacts to light and seem to validate the Paleoindian (but later than Clovis) age of at least some of these artifact forms. Work at Dust Cave is still in progress and the information presented here is preliminary, but present evidence shows the blades and blade tools to be concentrated in the lowest two levels of the site. The lesser proportion have been found in stratigraphic unit T, radiocarbon dated to ca. 10,000 radiocarbon years before present (RCYBP) and containing early side-notched points. More have been found in underlying stratum U, dated to ca. 10,300 RCYBP and containing an apparently mixed assemblage of Cumberland, Hardaway Dalton, and Quad points (Driskell 1994; Meeks 1994). Among the artifacts of interest are blades, tools made on blades, and tools made on bladelike flakes.

Eleven such artifacts examined in the course of this study illustrate the problems associated with retouched pieces that have a bladelike appearance. All of the eleven Dust Cave pieces are retouched and include two true blades, two flakes, and seven specimens on which the retouch is too invasive for the form of the original piece to be determined. Most, if not all, of these seven were probably fashioned from flakes (Fig. 8.12). One of the two true blades derives from an early Archaic provenience, and one from late Paleoindian. The early Archaic example (Fig. 8.12 b) is entirely different from known Clovis blades in that it is very thin and straight and has a large bulb and platform. The Paleoindian blade (Fig. 8.12 c) resembles some Clovis blades in cross section, curvature, and overall proportions; however, it also has a large bulb and platform. Both of these pieces were evidently detached using direct, hard-hammer percussion. Generally these do not compare very closely with known Clovis blades. The Dust Cave and other Middle Tennessee Valley (Hubbert n.d. and personal communication) specimens are less regular than Clovis blades and have large platforms, prominent bulbs, and moderately strong interior ripple marks indicating direct, hard-hammer percussion. Most of these blades are proportionally wider and thinner than Clovis blades. Prior blade scars on the exteriors are subparallel in most cases, and the overall configuration approaches that of large, narrow bifacial thinning flakes; however, I think that few if any of these were actually detached from bifaces.

No blade cores have been recognized thus far in the analysis of lithic specimens from Dust Cave, but Charles Hubbert (personal communication 1994) has recovered two interesting blade cores from the surfaces of sites near Dust Cave. One of these (Fig. 8.13 b, d), recovered from along Coffee Slough a few hundred meters from Dust Cave, has scars of two removals, one of which would be a flake and one of which would be a blade. Both would be thin and resemble bifacial thinning flakes, like the Dust Cave specimens. The core is plano-convex and was likely a large, primary cortex flake. Its convex surface is mostly cortical with marginal retouch, which constitutes preparation of acute platforms for detachment of large blades and flakes from the opposite, more planar face of the core. This piece is deeply patinated.

The second specimen is a true blade core (Fig. 8.13 c, e). It, too, was made on a large cortical flake, but in this case, the object blades were detached from the cortical face. There are four blade scars on the core face, one of which retains its deep negative bulb. Since this one resulted from the final blade removed from the core, the length and width of the blade can be determined from the scar to have been 104 mm and 35 mm, respectively. The platform is at an acute angle to the core face and was prepared with multiple, small flake scars (Fig. 8.13a).

Direct percussion would appear to have been used in the reduction of both of these cores. Neither of these cores has the configuration to produce the thicker blade from Stratum U of Dust Cave (Fig. 8.12c). The thickness of that blade indicates a polyhedral core with stronger ridges than occur on these two, so there is much more to be learned about blade technology in the latest Paleoindian to earliest Archaic manifestations of the Middle Tennessee Valley.

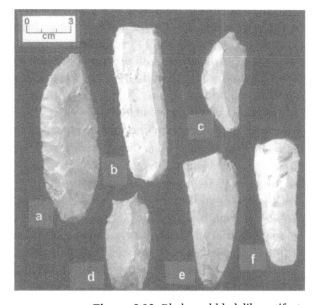

Figure 8.12. Blade and bladelike artifacts from Dust Cave, Alabama.

Figure 8.13. Cores found on the surface near Dust Cave, Alabama.

The Brookeen Creek Cache from central Texas consists of 173 lithic artifacts, mostly prismatic blades (Mallouf 1981, 1982). A majority of these were edge modified, and plow damage is inferred by Mallouf as responsible for most of this modification. Mallouf's two reports on this site are concerned with an analysis of the plow-damaged artifacts, and a final site report per se has not been published (additional information on the site is given in a survey report [Mallouf and Baskin 1976]). The cache was isolated and lacked any time-sensitive artifacts, thus leaving the temporal placement of the artifacts unresolved. Mallouf (1981, 1982) does not offer an interpretation, but Story (1990: 186) suggests that the Brookeen Cache blades may be of Clovis affiliation. Two factors lead me to believe that this cache is much younger. First, the morphology of the blades is not consistent with that of known Clovis blades. The Brookeen specimens are mostly wider, shorter, thinner, and less regular in outline; they have larger ripples on the interior surfaces and larger platforms and bulbs; they are less curved than Clovis blades; and, generally, they are minimally patinated. Second, the cache occurred in the upper 17 cm of valley fill deposits forming a terrace of Brookeen Creek (Mallouf 1982: 79, 82). The surface soil at this locality is Tinn Clay, subject to frequent flooding (Brooks 1978). Throughout the southern part of the United States, Clovis-age deposits are almost always found at or near the base of valley fill sections (Haynes 1992), an observation consistent with what is seen at such central Texas sites as Horn Shelter (Redder 1985; Story 1990; Watt 1978), Pavo Real (Henderson and Goode 1991), Wilson-Leonard (Collins et al. 1993; Holliday 1992, 1997; Masson and Collins 1995), Crockett Gardens (McCormick 1982), Kincaid (Collins 1990a; Collins et al. 1989), and Gault (Collins, Hester, and Headrick 1992). Tinn series soils are not easily assessed for age, but Brooks (1978: 155) classifies them as "Fine, montmorillonitic (calcareous), thermic Vertic Haplaquolls," neither especially young nor old. The attributes described above, together with recovery from a geomorphological context unlikely to be of Clovis age, is evidence that the Brookeen Cache is not of Clovis affiliation. As the comparative metric data show (Table 6.1; Figs. 8.5 and 8.10), these blades as an assemblage are wider and not as long as the Keven Davis blades, although there is overlap in both the length and the width ranges of the two groups; this is more apparent when thickness is also considered, because the Brookeen blades are also thinner. These characteristics match more closely those of some of the later blade assemblages such as Gibson and Weaver-Ramage than those of Clovis blades.

Seventy-two blades and modified blades found closely stacked in a cache pit in Coke County are inferred by Tunnell (1978) to be of

Archaic age. These generally have large platforms, prominent bulbs of percussion, high-amplitude ripple marks, and less regular overall form than is seen in the Clovis-age blades. The cache was found as it had been placed aboriginally, and, therefore, the modifications seen on the blade edges confidently can be inferred to represent intentional retouch. In fact, so many of the Gibson specimens are retouched that the cache might more accurately be referred to as a cache of scrapers. Only average metric data are available for the blades in this assemblage (Table 6.1).

Weaver-Ramage is another cache reported by Tunnell (1989), but in contrast to many lithic caches, this one contained a time-diagnostic arrowpoint, along with three 3 bifaces, 1 core, 8 tested cobbles, over 120 flakes, 24 unifacial tools, and 652 blades and blade segments (Tunnel 1989: 370). This cache was found in the Rolling Plains near the Salt Fork of the Brazos in western Texas. These blades belong to the Late Prehistoric Toyah interval, which is distinguished in part by the presence of blades. Blades in the Weaver-Ramage cache generally have large platforms, prominent bulbs, and irregular edges. The interior surfaces are strongly rippled, and, overall, the blades are smaller than Clovis blades. Tunnell provides only summary metric data (Table 6.1).

WEAVER-RAMAGE CACHE

Kirchmeyer is an open site near Corpus Christi Bay on the Texas Gulf Coast (Hester and Shafer 1975). Although the site has multiple components, Hester and Shafer infer that the blades and blade cores from the site are of Late Prehistoric affiliation. There are sixteen small blades (Table 6.1) of irregular outline with relatively large bulbs and platforms. The site locality is far from any sources of large chert pieces, which may partially account for the small size of the blades. These are obviously distinct from the Clovis blades in size and proportions (Figs. 8.4 and 8.9).

KIRCHMEYER

Indian Island is another site similar to Kirchmeyer, being an open, multiple-component campsite on the Texas coastal plain (Campbell 1956; Hester and Shafer 1975). A blade technology is present, inferred by Hester and Shafer to be of Late Prehistoric affinity. Sixteen small blades are documented by Hester and Shafer (1975) as having comparatively large bulbs and platforms with irregular edges. Again, the small size (Table 6.1) of these artifacts may be attributable in part to their distance from outcrops of larger chert pieces. These blades, too, differ in size (Fig. 8.4) and in proportions (Fig. 8.9) from Clovis blades.

INDIAN ISLAND

BARTON   The Barton site in central Texas is an open campsite and lithic workshop of Late Prehistoric (Toyah) affiliation (Ricklis 1995). Seven nearly complete blades from the site are typified by relatively large bulbs, large single-faceted platforms, and irregular edges. The blades are made of local chert that is of high quality and occurs as large nodules and stream cobbles. The large size (Fig. 6.1) of these blades in comparison with many of Late Prehistoric age is probably due in part to this favorable supply of chert. Some of the Late Prehistoric blades from the Barton site approach Clovis blades in size (Fig. 8.4) and proportion (Fig. 8.9).

MUSTANG BRANCH   At the Mustang Branch site is another Toyah component yielding blades (Ricklis 1995). Here were recovered at least three blade cores and seven blades in association with a large animal processing feature. These are large, irregular blades with comparatively large platforms and bulbs. Blade interiors have large ripples. Mustang Branch is in the same locality as the Barton site, and the large blades probably reflect the local abundance of good raw material (Table 6.1).

COMMENTS   From the foregoing comparative information, it is apparent that Clovis blades vary considerably but share several attributes in common. Blades from secure non-Clovis contexts can, for the most part, be distinguished. The Keven Davis blades in all respects fit well within the range of known Clovis blades. Therefore, the Keven Davis blades may be confidently inferred as being an isolated cache made by Clovis people. The significance of Clovis caching behavior is explored in the next chapters.

TIGHTLY CLUSTERED GROUPS OF ARTIFACTS IDENTI-
fied as Clovis have been reported (Frison 1991, Green 1963;
Hammatt 1970; Mallouf 1994), and are frequently referred to
as "caches," even though at least one clearly, and others possibly, represent burial goods that technically are not caches (Kornfeld,
Akoshima, and Frison 1990; Tunnell 1978). Among the better-known
Clovis "caches" (Table 3.1, Fig. 3.1) are Simon in southern Idaho
(Butler 1963, Butler and Fitzwater 1965; Woods and Titmus 1985),
Fenn probably in or near southwestern Wyoming (Frison 1991), Drake
in northeastern Colorado (Frison 1991; Stanford and Jodry 1988), at
least two at Blackwater Draw in east-central New Mexico (Green
1963; Montgomery and Dickenson 1992), Anadarko (McKee) in
southwestern Oklahoma (Hammatt 1970), and East Wenatchee in
central Washington (Mehringer 1988, 1989; Gramly 1993). Clovis artifacts at Anzick, south central Montana, were in (secondary?) burial
context (Frison 1991; Lahren and Bonnichsen 1974). Conceivably,
Fenn, Drake, Simon, or even East Wenatchee also were associated with
burials but evidence is lacking. Anzick, East Wenatchee, Fenn, and
Simon share traits in common—large Clovis points, preforms, ochre,
use of extraordinary raw materials (such as quartz crystal, obsidian,
and cherts exotic to the cache locality), and—at Anzick and East
Wenatchee—association with bone objects. Some of these cache
objects have traces of wear, indicating that they were not strictly ritualistic, but the properties just listed and the fact that the objects were
secreted in groups seems to remove them from purely utilitarian status.
The caches at Anadarko, Blackwater Draw, and Keven Davis, on the
other hand, contained prismatic blades of seemingly more ordinary,
utilitarian quality.

Five possible caches of Clovis artifacts have recently been noted in
Texas, at McFaddin Beach, in Bastrop County, at Crockett Gardens,

near Evant, and in San Antonio. Three Clovis points found in close proximity, though not in place, at the McFaddin Beach site are of nearly identical workmanship and material and are of extraordinary length compared to others known from the region (Hester et al. 1992: 21–22). These three points appear to be pristine and lack evidence for resharpening. C. K. Chandler (personal communication 1994) has recently documented two Clovis points found side-by-side on a surface exposure in Bastrop County ("unnamed site" in Table 3.1). The points are similar in workmanship, form, and raw material, and have been resharpened. Either of these groups of points could have come about as the result of some phenomenon other than caching, but caching should remain among the hypotheses considered whenever closely spaced artifacts are investigated and interpreted.

At the Crockett Gardens site in Williamson County, Texas, three Clovis blades were found in a small area of a single backhoe trench (McCormick 1982: 12.135–12.166), a concentration that may represent a disturbed cache. Elsewhere in the site were found two other blades and a mixture of other artifacts, including a Clovis point, lacking stratigraphic context.

Goode and Mallouf (1991: 67–70) describe three Clovis blade cores from near Evant in Hamilton County, which they believe may have been cached. It is also a distinct possibility that the two Van Autry Cores from Comanche Hill in Bexar County (Collins and Headrick 1992; Kelly 1992) were originally cached. All of these cores are of Edwards chert, and the Evant and Comanche Hill localities are both in areas where high-quality Edwards chert outcrops.

Caching is a distinctive behavior whereby people place "tools, equipment, and other materials in hiding or storage" as a "concrete expression of the anticipation of future needs" (Schlanger 1981: 4). The hidden materials are the cache, "an accumulation of useful material that is hidden away for future recovery and utilization. . . . [S]uch things as food, clothing, tools, and raw materials may be placed in a cache" (Tunnell 1978: 1). The implied distinction between caching and storage is that items are considered cached when the people go away and leave them, whereas storage occurs in or near an occupied habitation site. Obviously, what is storage during residence can become a cache upon abandonment of such a site.

The best evidence for Clovis-horizon caching is based on diagnostic artifacts, but food caching also has been postulated (Frison 1993; Frison and Todd 1986). In terms of adaptive strategies, caching of items of material culture (Binford 1979) is distinct from caching of food (Binford 1980, 1993), but both are important.

Food can be cached for short time periods as part of the strategy for moving it from source to point of consumption or for longer periods near residential sites for consumption during lean seasons. The distinction between "caching" and "storage" is less than sharp in the case of foods.

Objects of material culture may be cached for secular or for ritualistic purposes. Secular functions are associated with banking caches and abandonment caches, whereas ritual caching includes dedicatory and votive forms (Schiffer 1987: 79–80, 92–93). Banking caches are any form of secreting valuables for later recovery and use. This can occur systematically or as an ad hoc response to adversity. Banking is the apparent function of many caches, such as the lithic caches that commonly occur archeologically in Texas (Miller 1993; Tunnell 1978, 1989).

Abandonment caches are defined as consisting of utilitarian objects having less intrinsic value than those in banking caches, and as being left at sites to which a group anticipates returning. Probable examples of abandonment caches found archeologically at habitation sites in Texas are those of flakes and of grinding stones at the Salt Cedar site (Collins 1968) and manos at the Sleeper Site (L. Johnson 1991: 53–55, Fig. 22). Equipment cached near where its future use is anticipated, though not necessarily in habitation sites, is equivalent in function to abandonment caches. This situation can be inferred for hammerstones near quarries (Schiffer 1987: 93), duck decoys in caves near Humboldt Lake, Nevada (Heizer and Napton 1970), and wooden mortars and a pestle in the Lower Pecos region of Texas (Collins and Hester 1968; Prewitt 1981).

Dedicatory caches are those placed ritualistically into an architectural element (wall, floor, platform) of a structure as it is being built. Votive "caches" are accumulations of offerings left in or near a place of religious significance. Dedicatory caching occurs only in complex societies, whereas votive offerings may be found at any level of social complexity. Archeologically, in the absence of formal shrines, votive caches at what may be termed sacred places (Schiffer 1987: 80) can be difficult to distinguish from banking caches. Often the best clue is the nature of the objects cached, such as the split-twig figurines found in caches in the Southwest and thought to represent the invocation of hunting magic (Euler and Olson 1965; Jett 1958; Schwartz, Lange, and deSaussure 1958; Wheeler 1942).

Shawcross (1976) makes the case for votive artifact accumulations on the basis of accumulated broken objects in a spring in New Zealand. A similar suggestion for artifacts found accumulated in springs at Spring Lake in Texas was made by Shiner (1981) but later retracted (Shiner 1982). The presence of engraved stones in and near springs at the Gault site (Collins et al. 1991; Collins, Hester, and Headrick 1992; Hester, Collins, and Headrick 1992) might best be interpreted as votive. As Schiffer (1987: 80) observes, archeologists have recognized surprisingly few such occurrences given their prevalence in the ethnographic record.

Caches are important as archeological evidence for the logistical organization of hunter-gatherers (Binford 1979, 1980, 1993). Viewing caching behavior in somewhat different terms than Schiffer, Binford proposes that the function of what is cached should be the primary consideration. Food is cached by collectors as an integrated part of the subsistence behavior but almost never by foragers (Binford 1980). Objects of material culture subject to caching are of two kinds, in Binford's (1979) view. These are passive gear and insurance gear. Passive gear is seasonal in nature and is cached during the seasons when it is not needed. Insurance gear, in contrast, is not seasonal in nature but is cached "throughout the region, not in terms of specifically anticipated seasonal needs, but in terms of what might be generally needed at the location at some time in the future" (Binford 1979: 257). Passive gear is cached primarily by organized collectors, whereas both collectors and foragers cache insurance gear (Binford 1979, 1980; Gould 1980; Lee 1979).

It could be argued that the caches of duck decoys, milling stones, and wooden mortars and pestles mentioned above were seasonal equipment (Binford's passive gear). Nutting stones probably could be added to this list. Ducks, being migratory, probably were hunted only seasonally, and nutting stones probably were of use only when nuts ripened. Milling stones are less obviously seasonal, depending upon the range of plants milled and their seasons in a given habitat. Seeds found in the wooden mortar reported by Collins and Hester (1968) indicated that it had been used to process prickly pear tuna, a seasonal commodity. This would be a superb example of passive gear cached in anticipation of the next tuna harvest if it were certain that the mortar was not also used on other foods that, in aggregate, might be processed for most of the year (such as grass seeds, hardwood nuts, tule stalks, desert succulents).

Where do the many caches of stone artifacts fit into Binford's scheme? Caches of flakes, hammerstones and some other stone objects of general utility probably represent insurance gear; some could be passive gear left in areas of seasonal resource exploitation; still others could be banking caches related to exchange and long-distance transport apart from routine subsistence activities. Caches of biface pre-forms (e.g., Hart 1983; Hester and Brown 1988; Mallouf and Wulfkuhle 1991; Miller 1993; Witte 1942), cores (e.g., Goode and Mallouf 1991), flakes (e.g., Collins 1968; Green 1963: 150; Witte 1942), blades (Mallouf 1982), sets of tools (e.g., Brown 1985), projectile points (e.g., Janes 1930), and combinations of stone tools (e.g.,

Hart 1983; Slesick 1978; Tunnell 1978, 1989) in the south-central United States may be found isolated or within habitation sites; most of these remain of unknown purpose(s).

Additional careful and thorough study of caches of lithic artifacts is needed for a better understanding of their place in prehistoric adaptive strategies. Cached lithics should be examined for patterns of breakage and use wear as well as for the presence of chemical residues. Such information on their use history would contribute to understanding the purpose(s) of caching. The fact that most such caches are not found and investigated in situ by archeologists will probably continue to be a significant limitation to documentation of contexts, although it is likely that many caches, even if carefully excavated, would yield very limited contextual evidence of age or other aspects useful to interpretation of function.

TEN

Clovis Lithic Technology and Caching Behavior:
Clues to Subsistence Strategies and Behavior

C LOVIS ARTISANS FASHIONED TOOLS FROM HIGH-
quality stone using a versatile mix of flaking techniques and at
least two reductive strategies. In one strategy, bifaces were produced and reduced primarily into fluted "points." In the other, blades
were detached from polyhedral cores. Some of these blades were used
in an unmodified state as cutting tools. Various tools, such as end
scrapers, side scrapers, and burins, were made on other blades as well
as on flakes selected from the byproducts of sundry flaking activities,
such as preparing cores or making bifaces.

Among flakes produced by Clovis knappers were very large bladelike
pieces from which bifacial preforms, and ultimately points, were fashioned. Discard or loss of Clovis stone artifacts often occurred at great
distances from the geologic origin of the raw material. Durable Clovis
artifacts have been found cached at several localities and nondurable
goods, including food, may also have been cached at times. For example, there is good evidence for winter meat caching at the Sheaman
site in Wyoming (Frison 1982).

Known Clovis blades possess a distinctive constellation of attributes. They have small platforms in most cases, although some blades
have moderately wide (but never deep) striking surfaces. Clovis blades
are usually curved in longitudinal section. Bulbs and ripple marks on
the interior surfaces are relatively flat, giving a very smooth aspect to
this face. Generally, Clovis blades are long (frequently over 100 mm)
and narrow with robust cross sections, have relatively even and sharp
lateral margins, and have parallel to subparallel scars of previous
blades on the exterior surfaces. Length-to-width ratios almost always
exceed 3 to 1, commonly exceed 4 to 1, and even exceed 5 to 1 on occasion. Raw materials are always of high quality and commonly originate far from the point of archeological discovery.

The cores from which these blades were detached are of at least two

forms. The more common form has a large platform area nearly perpendicular to the long axis of the core. This platform surface is dim-
pled around its perimeter by small flake scars that intersect the core
face at acute angles. Apparently, the punch technique was used in the
detachment of blades from most cores of this form. The less common
kind of core has a smaller platform oriented at an acute angle to the
core face. These cores are best suited for direct-percussion reduction,
almost certainly done with a soft hammer. In some cases the "soft"
hammer seems to have been a chert stone with a thick cortex, a good
example of which was reported from the Yellow Hawk locality.

Clovis blade cores are almost never found with the negative bulbar
scars of the blades struck from them. This indicates that consistently
after each blade or each series of two or three blades was detached, the
platform was rejuvenated, which suggests preparation for additional
removals that never occurred. Even the very small (and one would
think depleted) core from the Clovis component at Kincaid Shelter
has a rejuvenated platform. That further reduction never occurred
on so many of these cores implies the lapse of an interval of time,
perhaps during which cores were stored or cached until fresh blades
were needed.

There is little evidence to suggest heat treatment of lithic raw materials by Clovis knappers. This is consistent with their favored use of
stone types that either do not improve with heating (quartz crystal,
obsidian) or that are of excellent quality without heat treatment (e.g.,
most varieties of Edwards chert).

A surprising degree of similarity is seen in both the bifacial and
blade aspects of Clovis lithic technology over the continent—enough
to imply considerable historical relatedness. The details of such a history are elusive at this juncture, but it seems plausible that Clovis
blade technology is ultimately derived from the often very similar ones
in Upper Paleolithic cultures of Eurasia, even though independent
invention is not an impossibility.

Of the Eurasian Upper Paleolithic blade technologies, the closest
similarities are in western European Aurignacian and Solutrean manifestations (Owen 1988; de Sonneville-Bordes 1960). The Solutrean of
France and Spain (ca. 21,000–19,000 B.P.) has long appealed to Bruce
Bradley (personal communications; see also Bradley quoted in Preston
1997: 76) as a candidate for the technological origins of Clovis biface
production because of the persistent use of overshot flaking in both.
With technological resemblances between Solutrean and Clovis
bifaces as well as between Solutrean and Clovis blade cores, one
wonders if Clovis knapping behavior may ultimately derive from western Europe. This possibility takes on even greater significance in light
of evidence of an apparently pre-Clovis biface- and prismatic-blade

technology from Meadowcroft Rockshelter and nearby sites in Pennsylvania (Adovasio 1993: 205–213). These pieces from Pennsylvania are diminutive compared with Clovis artifacts but are otherwise similar. The unfluted Miller biface from Meadowcroft (lower stratum IIa [between 11,300 ± 700 and 12,800 ± 870 B.P.]) exhibits what appear to be at least two overshot flake scars on one face (Adovasio 1993: Fig. 6 b) and, as illustrated, the associated prismatic blade technology inferred from Meadowcroft and related sites appears generally similar to those from the Upper Paleolithic of western Europe as well as to Clovis. Even the early human skeletal morphology from North America (Brace and Tracer 1992; Steele and Powell 1994) as well as South America (Neves and Pucciarelli 1991) could be interpreted to indicate origins either earlier than, or more western than, the Mongoloid populations of northeasternmost Asia.

We simply do not at this time know very much about social aspects of Clovis behavior, what the political landscape of the time was like, or precisely what environmental conditions prevailed and how Clovis people coped on a daily basis. It seems clear from the distribution of Clovis manifestations across most of the continent that environmental conditions were far from uniform, and likely the operative social units and their political interactions with other social units were equally variable. How much of the movement of tool stone across the landscape was by highly mobile groups and how much was through intergroup exchange of some kind? As Meltzer (1989) observes, the archeological end products of tool-stone acquisition directly from its source by the final user group are indistinguishable from those of acquisition indirectly via another, intermediary group, at our present level of archeological knowledge.

Eastern Texas is impoverished in tool stone suitable for virtually all characteristic forms of Clovis stone tools. Clovis artifacts found there either as isolated finds or at sites such as Aubrey, McFaddin Beach, or Keven Davis are predominantly of imported raw material, commonly Edwards chert. Outcrops along the eastern edge of the Edwards Plateau and gravel beds along streams flowing southeastward from the plateau are the closest sources for abundant, high-quality, large-sized pieces of chert.

"Edwards chert" is a catchall term for an array of chert varieties that outcrop over a wide expanse of Cretaceous uplands in central and west-central Texas (Banks 1990). These varieties range from coarse- to fine-grained, manifest many different colors, and have greater and lesser flaws or impurities that would be of concern to knappers. Most Edwards cherts do not require heat treatment, but some do; most will not tolerate heating. Most Clovis artifacts of Edwards chert in eastern Texas are made from the better varieties that do not require heating.

These varieties outcrop in restricted geographic localities for the most part. Also, there are important exceptions to the use of Edwards chert by Clovis peoples in eastern Texas.

There are Clovis artifacts made of fine-grained quartzite (Ferring 1989, 1990), unidentified cherts, and Manning fused glass (Brown 1994). It is clear from the use of multiple varieties of Edwards chert as well as the obscure Manning fused glass (Brown 1976) that Clovis people in eastern Texas were not just "coming into the country" as suggested by Kelly and Todd (1988). Rather, their knowledge of tool-stone sources was sophisticated and their means of acquiring stone afforded access to some of the best materials in the south-central part of the continent.

Looking at a wider geographic area, Clovis artifacts made of obsidian have been reported from a site on the Texas coast (Hester 1988) and at Kincaid Shelter (Hester et al. 1985) on the southwestern edge of the Edwards Plateau. The obsidian at Kincaid originated in the volcanic region of Queretero in central Mexico, nearly 1000 kilometers (600 miles) from the Kincaid site. One of the Clovis points from the Gault site (Collins et al. 1991) is made of Alibates chert from the Texas Panhandle (800 kilometers, or nearly 500 miles, away) and Chandler and McReynolds (1996: 7–9) report a fragmentary blade of Alibates material from San Patricio County on the Gulf Coastal Plain of Texas (over 930 kilometers, or 570 miles, from its source). I have examined the San Patricio specimen and believe it to be of Clovis origin. The Kincaid obsidian as well as the Gault and San Patricio County Alibates pieces are, respectively, broken and worn-out utilitarian pieces discarded at the locus of retooling. These are not objects of nonutilitarian exchange, although some added value or status may have been ascribed to pieces of exotic stone. At Clovis sites in central Texas where retooling occurred (Kincaid, Pavo Real, Yellow Hawk), discarded points are more commonly of local materials, and debitage is exclusively of local chert.

Evidence for the south-central United States suggests that at any given time a Clovis group would probably possess mostly tools made on stone acquired locally or from within a reasonably accessible distance (less than 200 kilometers), as well as a few made on stones from remote sources (such as the Alibates point at the Gault site and the obsidian points from Kincaid and the Gulf Coastal Plains site). All of the known specimens in both of these categories show evidence of having been used. From this pattern, it is tempting to suggest that Clovis knappers acquired chippable stone locally when it was available, stopped at known source areas as part of subsistence rounds, explored likely geologic exposures during subsistence wandering or any other travel into unfamiliar country, and exchanged stone opportunistically as part of social encounters whether scheduled or occurring by chance. Regrettably, as Meltzer (1989) has noted, elevating this notion from speculation to interpretation supported with unambiguous evidence is beyond the reach of our science at present. Much more detailed and intensive study of Clovis lithic artifact assemblages—including all forms of knapping debris—for patterns of breakage, wear and discard, loci of knapping, and refitting according to raw material categories would improve the data base and perhaps narrow the range of possible patterns of raw-material acquisition.

Caching of lithic stores in areas where tool stone is scarce, such as at the Keven Davis site, indicates anticipated return, probably as part of some form of subsistence round. Importantly, the numbers of Clovis caches in contrast to few or no Folsom caches is consistent with other evidence that indicates Folsom peoples were specialized bison-hunters and probably were less able to predict when they might be able to return to a cache. Clovis adaptations, on the other hand, are more of a foraging nature that likely involved more regular use of the landscape and a degree of predictable scheduling that made caching an effective strategy.

The Keven Davis blades were buried fairly deeply into the subsoil, which was enough to forestall weathering of the chert. Fresh Edwards chert noticeably alters in a matter of weeks or months if exposed to the elements, a fact surely known by any knapper with the knowledge to acquire this material and produce blades of the quality seen in this cache. It seems probable that beyond securing this cache from others, burial ensured that the chert pieces would remain fresh until retrieved—in this case, they remained pristine for probably 11,000 years.

Pavo Real and Adams are sites where considerable production of blades transpired, but blade-core reduction also occurred at Murray Springs, Gault, and probably other sites as well. At Pavo Real and at Adams, blades show considerable variation, and at Pavo Real many blades were broken, apparently during manufacture (at Adams, Blackwater Draw, and Keven Davis, most of the breakage was caused by machinery). The greater variability and breakage at the manufacturing sites suggests that the more uniform and complete blades were selected and carried away from the point of manufacture. The constellation of attributes found on those blades that were selected and removed from the workshops must reflect what Clovis knappers and users of the blades desired—those, then, are the true "Clovis Blades," with the constellation of attributes described above. Those attributes define a blade form that is almost as diagnostic of Clovis technology as are fluted Clovis points.

At workshops the debris, early blade removals, and rejected blades are more variable and include individual blades that could be confused with those of other technologies. At these workshops are also to be found large bladelike flakes that were struck from large blocks early in the preparation of blade cores or simply detached from pieces for which no further reduction was intended. These large, thick bladelike flakes are nearly straight in longitudinal section and were reduced bifacially into projectile point preforms (cf. Mallouf 1989: 99).

The smaller, curved, true blades were used as cutting tools, and segments of them were modified into scrapers, particularly end scrapers, and other forms of tools. The wear and bruising seen on ridges of blade cores and exteriors of blades suggest that cores were transported as raw material and blades struck off as needed. This inferred mobility is entirely consistent with the caching of blades and blade cores as insurance gear.

As more is learned about Clovis blade technology and its place in the broader range of Clovis behavior, some of the questions raised in this study may be answered. Of particular interest will be the origin of Clovis blade technology, why it seems to be more characteristic of Clovis manifestations in some regions of North America than in others, and the full range of its technological variations. Ultimately, it would be important to know why this technology was maintained in Clovis times but seems to have found little place in most later Paleoindian cultures.

When considering the origin of Clovis blade technology, one cannot escape the substantial similarities between Clovis blades and blade cores and those of the Upper Paleolithic from southwestern Europe (Owen 1988) to Siberia (Soffer 1985: 37–114). Among the Upper Paleolithic blades and blade tools illustrated by Soffer (1985: Figs. 2.46, 2.50, 2.78, 2.91, 2.107, 2.118, and 2.122) are numerous examples that qualify as microblades and microblade tools, "microlithization" in Soffer's (1985: 61, 218) words. Interestingly, microliths occur in small to moderate numbers throughout the Upper Paleolithic of the Central Russian Plain (at Mezin, ca. 29,700–21,600 B.P.; Yurovichi, ca. 26,470 B.P.; Khotylevo, ca. 24,950–23,600 B.P.; Mezhirich, ca. 19,280–14,300 B.P.; Timnovka I, ca. 15,300–12,200 B.P.; and Gontsy, ca. 13,400 B.P. [Soffer 1985: Table 2.12]). Microblades and microblade tools in the American Paleoarctic (ca. 10,600–9,000 B.P.) of Alaska (Dixon 1993: 51–68) are in all probability the easternmost and latest occurrences of the Siberian Upper Paleolithic. It seems highly probable that historical connections exist among all of these blade technologies, even if the full diversity of tools made on Upper Paleolithic blades in Europe and Russia are not found in American Paleoarctic and Clovis sites.

In this view, Clovis as it is generally defined is an arbitrarily bounded mix of archeological traits, none of which is completely unique to this "culture." Fluting of bifaces—"Clovis points"—may be the only technological innovation attributable to Clovis knappers. By and large, blade making seems to be a trait Clovis knappers inherited but failed to pass on to the cultures that succeeded them.

Postscript

I N LATE JULY AND AUGUST OF 1998, JUST AS THIS MAN-uscript was going to press, important finds relevant to Clovis blade technology came to light at the Gault Site in Bell County, Texas. The site is on a small Brazos River tributary that drains the eastern edge of the Lampasas Cut Plain in the central part of the state. The site has been previously reported as yielding Clovis-age engraved stones along with Clovis points, blades and blade cores (Collins et al. 1991; Collins et al. 1992; Hester et al. 1992). The owners of the land where the Gault Site is located contacted me at the Texas Archeological Research Laboratory (TARL) with the news that they had found a number of artifacts in the same geologic context as fossilized bones of large animals. T. R. Hester and I took a small group of staff and students from TARL and from the Vertebrate Paleontology Laboratory of The University of Texas at Austin to the site to jacket and remove a block of matrix containing some of the bones. Later we conducted a few days of exploratory excavations at the locality, and have now begun to document the numerous artifacts recovered in TARL's excavation and that of the landowners. Further controlled work at this site is underway and more is planned for the future.

The Gault Site occupies a verdant cove at the head of a small valley cut into Edwards Limestone. Numerous springs seep and flow from the limestone and large nodules of the finest Edwards chert are abundant at and near the locality. Large quantities of Edwards chert flaking debris and artifacts are present. A dark midden of Holocene age overlies a complex suite of deposits of Late Pleistocene age. These Pleistocene deposits are a rocky colluvium, lenses of ponded clay, and an intricate network of spring throats along the margin of the valley. The colluvium intrudes into the springs and has been reworked and modified by the action of groundwater. It is the bones and selected lithic artifacts from these older deposits that are briefly and preliminarily described here.

Among the mostly fragmentary bones and teeth were elements definitely identifiable as belonging to bison, horse, and mammoth. The block contained the ulna of a (juvenile?) mammoth and other fragments of proboscidean-size bones; *in situ* with these bones were chert flakes and a Clovis point. The bone is fossilized but not in very good condition. On a later date, the almost complete mandible of a juvenile mammoth was found and it, too, was associated with numerous flakes.

The deposit yielding the fossil bone and the *in situ* Clovis point is iron-stained to a yellowish color as are all chert pieces from that deposit. This characteristic staining allows the inference to be made with reasonable confidence that certain dislodged artifacts are derived from the older deposits. Among the recently exposed pieces inferred to be from the stratum with *in situ* Clovis points and bones of extinct animals are the blade cores, blades, and blade tools described below. Before discussing the blade technology, mention is made of several stained artifacts resulting from biface technology and inferred to be of Clovis affiliation. This discussion is preliminary since analysis of these materials has barely begun, but even at this stage of study, the blade data add significantly to the interpretations offered in the preceding volume.

Clovis biface technology is indicated by 4 points, 1 point fragment, 6 unbroken preforms, 17 preforms probably broken during manufacture (including 2 broken by diving flutes), a channel flake and other debitage. Among the debitage are several overshot flakes, one of which was detached from a biface over 130 mm wide. Other than the size of the one very wide biface, these materials are comparable to Clovis bifaces from several sites on and near the southern periphery of the Great Plains.

Thirteen blade cores and a fragment have been recovered. Only one of the 13 could be called a conical core, and it is small and atypical. The general form of the fragmentary core is problematical. Twelve are wedge-shaped cores with acute angles between the core platform and face. Four of these 12 are flattish pieces unlike any previously documented from Clovis sites. The other 8 (Postscript Fig. 1 a–c) have strongly convex core faces and closely resemble a majority of the experimental blade cores produced by Glenn Goode (c.f. Figs. 2.3, 2.4, and 2.10 a, c, d, g, and h). This represents a much higher incidence of wedge-shaped Clovis cores than had been observed at the time this book was written. It is perhaps significant that the best example of the wedge-shaped cores discussed in Chapter 3 was from the Gault Site (see Fig. 3.13 a and d). Two of the 8 wedge-shaped cores have remnants of a bifacial crest at the distal end of the core (Postcript Fig. 1 a). Several of these cores have blade scars with the negative bulb intact, indicating less frequent removal of the platforms than is seen on the conical cores (Postcript Fig. 1 c). The final blade scars on the faces of the 8 wedge-shaped cores measure between 75 and 103 mm in length.

There is one flat, thin, oval, multifaceted flake that is tentatively identified as an atypical core tablet. It may have been removed from the platform of a wedge-shaped core as it is almost identical to one produced experimentally by Glenn Goode (Fig. 2.10 a). Although it has not been systematically examined, the abundant collection of small debitage contains numerous examples of blade fragments and other debris characteristic of blade production.

Two specimens in this collection (one complete and one fragmentary) consist of unsuccessful blade detachments, one diving and the other misdirected into the interior of the core. These are neither cores nor blades, strictly speaking, since they each consist of a blade and a substantial portion of core in a single piece.

The recently collected materials include 13 complete blades and fragments of 36 more (Postcript Fig. 1 d–e). These cover the entire range from largely cortical (stage 1) to advanced, non-cortical (stages 5 and 6) pieces; numerous examples of both stages 3 and 4 are present. These 49 specimens exhibit no clear evidence of intentional edge modification nor of macroscopic use-damage. Several of the proximal fragments show large eraillure scars or platform crushing. These may be symptomatic of removing blades with direct percussion. They have not been examined microscopically. Lengths range up to 148 mm.

Another group of unmodified blades are crested (stage 2). Three of these have complete crests (Postcript Fig. 1 d) and two crested blade fragments seem to have been the first removals from a crested core face. Seven additional blades (Postcript Fig. 1 e) and one fragment have partial crests.

Blades showing signs of use are in two groups at the present stage of analysis. The first group includes 10 blades and 4 fragments with edge damage visible to the unaided eye, consisting of nicking and microflaking along one or both edges. Two other blades have been examined microscopically and found to have well developed use wear. One secondary cortical (stage 1) blade (Postcript Fig. 1 g) has bright "sickle sheen" along one edge and the adjacent faces. This sheen is lightly striated parallel to the long axis of the blade. This would appear to be a tool used in cutting plant material rich in biosilicate. The other blade has patchy dull polish on both faces and multidirectional striations; no specific contact material nor use pattern has been inferred from the use wear on this specimen. Time has permitted the microscopic examination of only three pieces in this collection (these two and the small serrated blade discussed below) and all three show signs of use.

Postscript Figure 1. Blade cores, blades, and blade tools recovered from the Gault Site (41BL323) in July and August, 1998. Wedge-shaped cores (a–c), fully crested blade (d), partially crested blade with marginal retouch (e), retouched blade (f), blade with "sickle sheen" along the edge (g), dihedral burins on blades (h, i), burin-on-truncation and dihedral burin (j), enlarged view of central portion of large, serrated blade (k).

Of particular interest are several tools made on blades. One large blade (138 mm long) is deeply serrated along both edges (Postscript Fig. 1 k). There are a smaller serrated blade and three fragments. One of these fragments has been examined microscopically and found to have very slight polish and well-developed, fine striations oriented diagonally to the edge. The striae probably result from cutting some soft material, such as meat. Unifacial retouch appears along the margins of several blades (Postscript 1 e, f). Finally, there are two dihedral burins on blades (Postscript Fig. 1 h, i) and a complex burin (Postscript Fig. 1 j). The latter is a heavily retouched blade (or possibly a flake) with a concave truncation at the proximal end from which a spall was struck that detached along the entire length of the piece; a second spall was detached from this distal termination to form a dihedral burin. Overall, the piece has a burin-on-truncation at one end and a dihedral burin at the opposite end.

Several tentative conclusions are suggested by the preliminary examination of this assemblage. First it is apparent that Clovis knappers made large numbers of bifaces as well as blades at the Gault site. This is not surprising given the abundance of quality raw material. It further appears that the preferred approach to blade making was to produce and reduce wedge-shaped cores. These cores often had completely chipped crests to guide the first blade removal. Platform renewal is required less often on wedge-shaped than on conical cores. The impression I gained early in my study of Clovis blades was that indirect percussion was the most commonly used technique for blade making. Several colleagues looking at my data, and especially Glenn Goode and others viewing the evidence from the perspective of knappers, have expressed the opinion that direct percussion may have been more common than I thought. Certainly the overall character of the present assemblage suggests direct percussion, and my views are tending to shift toward closer agreement with those who see Clovis blades as mostly the product of direct percussion.

Two sites, Gault and Pavo Real, have produced a preponderance of wedge-shaped cores, but two conical cores are known from Gault and a discoidal core tablet is among the specimens from Pavo Real. The significance of the differences between these two core forms is unknown at the present time. That both occur at the same sites made out of local materials may negate any suggestion that the differences are in response to raw materials. Temporal differences cannot be ruled out with the present lack of a chronology within the Clovis interval. And, of course, other possibilities exist, such as ethnic differences, availability of suitable billets or punches, immediacy of need, and others.

Numbers of retouched blades and burins on blades at Gault are comparatively higher than in most assemblages. Also, the very limited amount of microscopic examination indicates that blades from this site will yield valuable data on blade use. Overall, preliminary indications are that this assemblage shows greater modification of blades into tools and greater use of unmodified blades than had been discerned from the several previously studied examples and assemblages. Clearly, the Gault site has much to offer in the continued study of early Paleoindian culture.

I am extremely grateful to the Lindsey family, Howard, Doris, Ricky, and Leslie, for allowing investigations on their property and for facilitating study of their collection by TARL staff and students. Also, I thank Dale Hudler of TARL's microscopy laboratory for taking a day out of his busy schedule to examine three blades from the Gault site, Milton Bell for the photographs used in the accompanying figure, and Frank A. Weir for pasting-up that figure.

ANGLE OF FLAKING: the inclination, measurable in degrees, between the plane of a striking platform and the direction, or vector, or force that is applied in the detachment of a flake or blade whether that force be pressure, direct percussion, or indirect percussion; the angle of flaking is almost always less than 90 degrees.

ANGLE OF FORCE: *See* angle of flaking.

ARRIS: the ridge or crest formed by the intersection of two flaked planes or facets, either on the face of a core or on the exterior of a flake or blade.

BACKED BLADE: a blade with one dull edge, whether dulled by flaking or by the presence of cortex (referred to as naturally backed or cortically backed), opposite a sharp edge.

BATON: an elongate, generally cylindrical percussor of wood, bone, or antler used in soft-hammer flaking; also referred to as a billet or as a soft hammer.

BEVEL: an inclined plane formed by multiple flake scars along the edge of a piece of knapped stone; beveling is commonly employed to form the working edge of a tool or to serve as a platform for subsequent flake removals (usually on the face opposite the bevel).

BIFACE: a piece of knapped stone with flake scars on both faces.

BIFACIAL: adjective indicating the presence of flake scars on both faces of the modified noun.

BILLET: *See* baton.

BLADE: a specialized flake removed from a prepared core; the flake is at least twice as long as it is wide and exhibits parallel to subparallel blade scars on its exterior surface (see Chapter 2 for an extended discussion).

BLADE TOOL: any chipped-stone tool produced by the modification of a blade.

BLANK: a partially formed, chipped-stone tool suitable for further flaking to completion; in some usages, the term is applied very narrowly and is distinguished from *preform,* but these usages vary significantly and one scholar's blank is another's preform.

BULB: the raised and rounded eminence on the proximal, ventral surface of a flake or blade, just distal of the striking platform.

BULB OF PERCUSSION: the rounded eminence on the proximal interior surface of a flake or blade.

BULB SCAR: *See* erraillure scar.

BULBAR SCAR: the rounded depression near the proximal end of a flake or blade scar on a core or other piece of knapped stone.

BURIN: a specialized, chisel-like tool, the working edge of which is formed by the acute intersection of two planes, at least one (but usually both) of which was formed by the removal of a prismatic flake (the burin spall).

CHAPEAU DE GENDARME: a term used to describe the characteristic recurved shape of the heavily retouched platforms on sequent blades and points removed from Levallois cores, typical of certain Old World Middle Paleolithic cultures.

CHIP: small, thin fragment of a flake lacking the bulb and platform.

COBBLE: a piece of stone either rounded by stream action or having its maximum diameter between 6.4 and 25.6 cm, or both.

COMPOUND DART: a projectile consisting of multiple parts, including some combination of main shaft, fore shaft, one-piece dart point, fletching, or composite point (formed by multiple microlithic pieces set in sockets to serve as tip, cutting edges, and barbs).

CONCHOIDAL: a term descriptive of the manner in which isotropic materials fracture, the detached piece having a smooth, slightly rippled, convex exterior resembling the outer surface of a clam shell; the scar left on the parent piece bears the negative image of these attributes and resembles the interior of a clam shell.

CORE: a mass of raw material from which flakes, blades, or bladelets have been detached.

CORE TOOL: a tool made on a core, usually by the patterned removal of flakes.

CORTICAL: adjective referring to the presence of cortex on the object described, as a cortical blade.

CORTEX: altered or weathered exterior of siliceous stone, usually lacking in the desirable properties of hardness, sharpness of edge, and chippability found in the unweathered interior; most knapping begins with the systematic removal of all or most cortex.

CREST: the ridge formed by flaking (usually bifacially) the face of a core; this ridge serves as a guide for the removal of blades or bladelets.

CRESTED: refers to the presence of a flaked crest on the face of a core or the remnant of such a crest on a blade.

DEBITAGE: a term borrowed from the French and used in English as a noun to refer to all of the byproducts of knapping, especially flakes, but also blades, cores, chunks, chips, shatter, and debris; in French, the term is a verb referring to the act of knapping.

DEBRIS: nondescript, irregular products of knapping, usually thought of as having little or no utility to the knapper.

DIRECT PERCUSSION: the technique of flaking in which flakes are detached by striking the parent piece with a percussor; also sometimes used to describe block-on-block flaking, in which the core is struck against a fixed block, or "anvil."

DISTAL END: the end opposite the origination of force on a core, flake, or blade.

ERRAILLURE SCAR: a tiny, often nearly triangular flake scar on a bulb of percussion on the interior of a flake or blade; sometimes called a bulb scar, a term that may cause confusion with bulbar scar.

EXTERIOR SURFACE: the surface of a flake or blade that existed prior to detachment of that flake or blade from the core.

FISSURE: irregular and often discontinuous cracks or furrows on the interior surface of a flake or blade or on flake or blade scars on the face of a core; fissures often radiate outward from the point of force application.

FLAKE: a piece of siliceous material bearing the attributes of a bulb of percussion and a platform produced by the removal of that piece from a core or other parent piece.

FLAKE SCAR: the flattish to slightly concave surface left on a core or other parent piece by the removal of a flake.

FLAKE TOOL: a tool made on a flake, usually by the patterned removal of flakes.

FLUTE: an elongate flake scar on the face of certain types of projectile points, usually emanating from, and resulting in, the thinning of the base (examples of fluted point types include Clovis, Cumberland, Folsom, Fells Cave Stemmed, and Gainey); also referred to as a channel scar.

HAMMER: any object used to strike a piece of siliceous stone for the removal of flakes.

HAMMERSTONE: a piece of stone used by a knapper as a hammer for the flaking of stone; usually distinguished by the presence of battered and scarred surfaces.

HARD HAMMER: any percussor of very durable material, but usually of stone, used in knapping; the products of hard-hammer flaking generally include flakes with prominent bulbs of percussion and large platforms.

HINGE FRACTURE: a detachment in which the force fails to carry to a thin, tapered termination, leaving a rounded distal edge (convex) on the flake and a rounded (concave), steplike scar on the parent piece; generally thought of as a knapping error.

INDIRECT PERCUSSION: technique of flaking in which an object, usually of antler or bone and referred to as a punch, is interposed between the percussor and the core; the punch is held with one end firmly in contact with the core and the opposite end of the punch is struck; this arrangement allows for precise control of the angle of flaking and the point at which force enters the core.

INTERIOR SURFACE: the surface of a blade or flake that was formed along the plane of detachment from the parent piece.

INVASIVE RETOUCH: flaking that extends a substantial distance onto the face of the parent piece; flake scars that are not confined to a narrow margin along the edge of the parent piece.

KNAP: the act of controlled fracturing of siliceous stone; the work is done by a knapper; the product is said to be knapped.

LAMELLAR: French adjective meaning bladelike.

LANCEOLATE: an adjective referring to the shape of chipped stone pieces that have convex edges tapering to a point at one or both ends.

LEFT MARGIN: the edge of a blade to the observer's left when looking from the proximal (platform) end toward the distal end with the exterior surface up.

LIP: a small protrusion or overhang on the face of a core produced by the intersection of the core platform and the negative bulbar scar of a detached flake or blade; a small protrusion or overhang on the ventral/proximal end of a flake or blade, commonly produced in soft-hammer percussion.

MACROBLADE: a term used by some to refer to very large blades (typically greater than 15 cm in length) but by others to distinguish ordinary-sized blades from microblades; in this study, the term is avoided because of this ambiguity, but its usage to refer to extraordinarily large blades is preferred.

MICROBLADE: small blades for which various definitions have been proposed; used in this study to refer to blades generally less than 3 cm long and 1 cm wide.

NEGATIVE: the scar or depression left on a parent piece by the removal of a blade or flake.

NODULE: an irregular, rounded mass of siliceous stone found in, or weathered from, the host bedrock.

OVERSHOT: fracture in which a flake or blade carries completely across the face of the parent piece and removes a portion of the edge opposite the platform; common in Clovis bifacial flaking.

PATINA: weathered or chemically altered surface of siliceous stone.

PATINATION: the process by which the surface of sileceous stone weathers or is chemically altered; often a slow process.

PERCUSSION: direct impact of a moving percussor against a core or parent piece or of a moving core against a fixed block or anvil.

PERCUSSOR: any object used to deliver force during knapping; includes hammerstones, billets, batons; the percussor may strike the core directly or deliver force indirectly through a punch; synonymous with hammer.

PERVERSE FRACTURE: a helical, twisting, or spiral break initiated at the point of force along the edge of the piece being flaked; the plane of flaking begins subparallel to the face of the parent piece and twists toward a perpendicular orientation and causes the parent piece to break.

PLATFORM: the surface area to which force is delivered in the removal of a flake or a blade; on the parent piece, a complete platform exists prior to removal, and a reduced portion exists after removal; the platform is the proximal edge of a detached flake or blade; platforms may be prepared or natural.

PREFORM: a partially formed, chipped-stone tool suitable for further flaking to completion; *see also* blank.

PRESSURE FLAKING: a technique for working stone in which force is loaded directly by placing a tool against the platform and increased until fracture occurs.

PRIMARY FLAKE: a piece removed at the beginning of the reduction of a mass of raw material; generally distinguished by cortex completely or nearly completely covering the exterior of the flake.

PROXIMAL END: the end of a core or a flake at which force of detachment originated.

REJUVENATION: an operation that restores the platform of a core to usable form after flake or blade removals had rendered it unsuitable for continued reduction; may be accomplished by removing a single large flake or several smaller flakes.

RETOUCH: one or more controlled flake detachments for the purpose of shaping the face or edge of a chipped-stone object.

RIDGE: an elongate intersection of two planes on the face of a core or the exterior of a flake or blade; the planes may consist of single or multiple facets.

RIGHT MARGIN: the edge of a blade to the observer's right when looking from the proximal (platform) end toward the distal end with the exterior surface up.

RIPPLES: a succession of rounded, wavelike ridges and swales on the fracture-plane surface of chipped stone; ripples are concave toward the point of force initiation.

SECONDARY FLAKE: a piece removed at an intermediate stage in the reduction of a mass of raw material; the flake exterior is partially covered in cortex but also exhibits scars from the earlier removal of flakes.

SHATTER: irregular and often thick and angular pieces of stone produced by knapping but lacking the diagnostic attributes of flakes.

SOFT HAMMER: any percussor softer than the siliceous material being flaked; often used synonymously with billet or baton in reference to hammers of antler, bone, or wood, but can also denote a hammer of soft stone.

STEP FRACTURE: a flake or flake-scar termination that forms a flat surface at nearly a right angle to the long axis of the flake; actually a break that occurs in the flake being removed leaving a partially detached mass on the face of the parent piece; multiple step fractures often occur, leaving a stack or knot on the face of the parent piece; generally considered to be a knapping error.

STRIKING PLATFORM: the surface of a flake or a blade in contact with the punch, pressure flaker, or percussor at the time force of detachment is applied; a portion of the striking platform remains on the parent piece and a portion detaches with the flake or blade being removed.

TOOL STONE: any rock or mineral suitable as raw material for the manufacture of stone tools; generally used to refer to commonly used varieties.

UNIFACE: a piece of knapped stone with flake scars on only one face.

VENTRAL: the face of a flake or blade formed by the fracture plane at the time the piece was removed from its parent mass; the interior surface.

ADAIR, L.C. 1976. The Sims Site: Implications for Paleoindian Occupation. *American Antiquity* 41: 325–334.

ADOVASIO, J.M. 1993. The Ones That Will Not Go Away: A Biased View of Pre-Clovis Populations in the New World. In *From Kostenki to Clovis,* edited by O. Soffer and N. D. Praslov, pp. 199–217. Plenum, New York.

ADOVASIO, J.M., J. D. Gunn, J. Donahue, and R. Stuckenrath 1977. Progress Report on the Meadowcroft Rockshelter—A 16,000 Year Chronicle. In *Amerinds and the Paleoenvironments in Northeastern North America,* edited by W. S. Newman and B. Salwen, pp. 137–159. Annals of the New York Academy of Sciences 288.

AHLER, S.A., and R. B. McMillan 1976. Material Culture at Rodgers Shelter: A Reflection of Past Human Activities. In *Prehistoric Man and His Environments,* edited by W. R. Wood and R. B. McMillan, pp. 163–200. Academic Press, New York.

BANKS, L.D. 1990. *From Mountain Peaks to Alligator Stomachs: A Review of Lithic Sources in the Trans-Mississippi South, the Southern Plains, and Adjacent Southwest.* Oklahoma Anthropological Society, Memoir 4.

BAMFORTH, D.B. 1988. Investigating Microwear Polishes with Blind Tests: The Institute Results in Context. *Journal of Archaeological Sciences* 15: 11–23.

BARNES, V.E. 1988. Dallas Sheet. *Geologic Atlas of Texas.* Bureau of Economic Geology, The University of Texas at Austin.

BAUMLER, M.F., and C. E. Downum 1989. Between Micro and Macro: A Study in the Interpretation of Small-Sized Lithic Debitage. In *Experiments in Lithic Technology,* edited by D. S. Amick and R. P. Mauldin, pp. 202–216. BAR International Series 528. Oxford.

BECKER, M.J. 1973. Archaeological Evidence for Occupational Specialization among Classic Period Maya at Tikal, Guatemala. *American Antiquity* 38: 396–406.

BEMENT, L.C. 1986. *Excavation of the Late Pleistocene Deposits of Bonfire Shelter, Val Verde County, Texas.* Texas Archeological Survey, Archeology Series 1. The University of Texas at Austin.

BINFORD, L.R. 1979. Organization and Formation Processes: Looking at Curated Technologies. *Journal of Anthropological Research* 35: 255–273.

1980. Willow Smoke and Dog's Tails: Hunter Gatherer Settlement Systems and Archaeological Site Formation. *American Antiquity* 45: 4–20.

1993. Bones for Stones: Considerations of Analogues for Features Found on the Central Russian Plain. In *From Kostenki to Clovis,* edited by O. Soffer and N. D. Praslov, pp. 101–124. Plenum, New York.

BOKSENBAUM, M.W., P. Tolstoy, G. Harbottle, J. Kimberlin, and M. Neivens 1987. Obsidian Industries and Cultural Evolution in the Basin of Mexico before 500 B.C. *Journal of Field Archeology* 14: 65–76.

BORDAZ, J. 1970. *Tools of the Old and New Stone Age.* Natural History Press, Garden City.

BORDES, F. 1947. Etude comparative des differentes techniques de taille du silex et des roches dures. *L'Anthropologie* 51: 1–29.

1953. Notules de typologie Paleolithique, II: Pointes Levalloisiennes et pointes pseudo-Levalloisiennes. *Bulletin de la Societe Prehistorique Français* 50: 311–313.

1961. *Typologie du Paleolithique ancien et moyen.* Delmas, Bordeaux.

1967. Considerations sur la typologie et les techniques dans le Paleolithique. *Quartar* 18: 25–55.

1968. *The Old Stone Age.* McGraw-Hill, New York.

BORDES, F., and D. Crabtree 1969. The Corbiac Blade Technique and Other Experiments. *Tebiwa* 12 (2): 1–21.

BRACE, C.L., and D. P. Tracer 1992. Craniofacial Continuity and Change: A Comparison of Late Pleistocene and Recent Europe and Asia. In *The Evolution and Dispersal of Modern Humans in Asia,* edited by T. Akazawa et al., pp. 439–472. Hokusensha Publishing, Tokyo.

BRADLEY, B.A. 1982. Flaked Stone Technology and Typology. In *The Agate Basin Site,* by G. C. Frison and D. J. Stanford, pp. 181–208. Academic Press, New York.

1993. Paleo-Indian Flaked Stone Technology in the North American High Plains. In *From Kostenki to Clovis,* edited by O. Soffer and N. D. Praslov, pp. 251–262. Plenum Press, New York.

BRENNAN, L.A. (EDITOR) 1982. A Compilation of Fluted Points of Eastern North America by Count and Distribution: An AENA Project. *Archaeology of Eastern North America* 10: 27–46.

BREZILLON, M. 1968. La denomination des objets de pierre tailee: Materiaux pour un vocabulaire des prehistoriens de langue Française. *Gallia Prehistoire,* supplement 4. Centre National de la Recherche Scientifique, Paris.

BROOKS, C.A. 1978. *Soil Survey of Hill County, Texas.* Soil Conservation Service, United States Department of Agriculture, Washington, D.C.

BROSTER, J.B., and M. R. Norton 1993. The Carson-Conn-Short Site (40BN190): An Extensive Clovis Habitation in Benton County, Tennessee. *Current Research in the Pleistocene* 10: 3–4.

BROWN, K.M. 1976. Fused Volcanic Glass from the Manning Formation. *Bulletin of the Texas Archeological Society* 47:189–207.

1985. Three Caches of Guadalupe Tools from South Texas. *Bulletin of the Texas Archeological Society* 56: 75–126.

1994. Four Clovis Points from San Augustine County, Texas. *La Tierra* 21 (2): 24–39.

BRYAN, W.A., and M. B. Collins 1988. *Inventory and Assessment of Cultural Resources Above the 1551.5-foot Contour Line, Stacy Reservoir Recreation Areas, Concho, Coleman and Runnels Counties, Texas.* Prewitt and Associates, Inc., Reports of Investigations 57. Austin.

BUTLER, B.R. 1963. An Early Man Site at Big Camas Prairie, South-Central Idaho. *Tebiwa* 6 (1): 22–23.

BUTLER, B.R., and R. J. Fitzwater 1965. A Further Note on the Clovis Site at Big Camas Prairie, South-Central Idaho. *Tebiwa* 8 (1): 38–40.

BYERS, D. 1954. Bull Brook—A Fluted Point Site in Ipswich, Massachusetts. *American Antiquity* 19: 343–351.

1966. The Debert Archaeological Project: the Position of Debert with Respect to the Paleo-Indian Tradition. *Quaternaria* 8: 33–47.

CALLAHAN, E. 1979. The Basics of Biface Knapping in the Eastern Fluted Point Tradition: A Manual for Flintknappers and Lithic Analysts. *Archaeology of Eastern North America* 7 (1): 1–180.

CAMPBELL, T.N. 1956. Archeological Materials from Five Islands in the Laguna Madre, Texas Coast. *Bulletin of the Texas Archeological Society* 27: 7–46.

CARLSON, D.L., and D. G. Steele 1992. Human-Mammoth Sites: Problems and Prospects. In *Proboscidean and Paleoindian interactions,* edited by J. W. Fox, C. B. Smith, and K. T. Wilkins. Baylor University Press, Waco.

CHANDLER, C.K. 1992. A Polyhedral Blade Core from Northeast San Antonio, Bexar County, Texas. *La Tierra* 19 (4): 20–25.

CHANDLER, C.K., and R. McReynolds 1996. Artifacts of Alibates Dolomite from South Texas. *La Tierra* 23 (3): 7–9.

CHARLTON, T.H. 1978. Teotihuacan, Tepeapulco, and Obsidian Exploitation. *Science* 200: 1227–1236.

CLARK, D.W. 1991. The Northern (Alaska-Yukon) Fluted Points. In *Clovis: Origins and Adaptations,* edited by R. Bonnichsen and K. L. Turnmire, pp. 35–48. Center for the Study of the First Americans, Department of Anthropology, Oregon State University, Corvallis.

CLARK, J.E. 1982. Manufacture of Mesoamerican Prismatic Blades: An Alternative Technique. *American Antiquity* 47: 355–376.

1984. Counterflaking and the Manufacture of Mesoamerican Prismatic Blades. *Lithic Technology* 13 (2): 52–61.

1985. Platforms, Bits, Punches, and Vises: A Potpourri of Mesoamerican Blade Technology. *Lithic Technology* 14 (1): 1–15.

1987. Politics, Prismatic Blades, and Mesoamerican Civilization. In *The Organization of Core Technology,* edited by J. K. Johnson and C. A. Morrow, pp. 259–284. Westview Press, Boulder.

1989. La fabricacion de navajas prismaticas. In *La Obsidiana en Mesoamerica,* edited by M. Gaxiola and J. E. Clark, pp. 145–155. Serie Arqueologia, Instituto Nacional de Antropologia e Historia, Mexico City.

CLARKE, D.L. 1968. *Analytical Archaeology.* Methuen, London.

CLARKE, R. 1935. The Flint-Knapping Industry at Brandon. *Antiquity* 9: 38–56.

COE, M.D. 1994. *Mexico from the Olmecs to the Aztecs.* Thames and Hudson, London.

COLLINS, M.B. 1968. *The Andrews Lake Locality: New Archeological Data from the Southern Llano Estacado.* unpublished master's thesis, The University of Texas at Austin.

1990a. The Archeological Sequence at Kincaid Rockshelter, Uvalde County, Texas. *Transactions of the 25th Regional Archeological Symposium for Southeastern New Mexico and Western Texas,* pp. 25–33. Midland Archeological Society, Midland.

1990b. Observations on Clovis Lithic Technology. *Current Research in the Pleistocene* 7: 73–74.

1996. The Keven Davis Cache (41NV659) and Clovis Blade Technology in the South Central United States. Report on file, Office of the State Archeologist, Archeology Division, Texas Historical Commission, Austin.

COLLINS, M.B., M. D. Blum, R. A. Ricklis, and S. Valastro 1990. Quaternary Geology and Prehistory of the Vara Daniels Site, Travis County, Texas. *Current Research in the Pleistocene* 7: 8–10.

COLLINS, M.B., C. B. Bousman, P. Goldberg, P. R. Takac, J. C. Guy, J. L. Lanata, T. W. Stafford, and V. T. Holliday 1993. The Paleoindian Sequence at the Wilson-Leonard Site, Texas. *Current Research in the Pleistocene* 10: 10–11.

COLLINS, M.B., G. L. Evans, T. N. Campbell, M. C. Winans, and C. E. Mear 1989. Clovis Occupation at Kincaid Shelter, Texas. *Current Research in the Pleistocene* 6: 3–4.

COLLINS, M.B., and P. Headrick 1992. Comments on Kelly's Interpretations of the "Van Autry" Cores. *La Tierra* 19 (4): 26–39.

COLLINS, M.B., and T. R. Hester 1968. A Wooden Mortar and Pestle from Val Verde County, Texas. *Bulletin of the Texas Archeological Society* 39: 1–8.

COLLINS, M.B., T. R. Hester, and P. J. Headrick 1992. Engraved Cobbles from the Gault Site, Central Texas. *Current Research in the Pleistocene* 9: 3–4.

COLLINS, M.B., T. R. Hester, D. Olmstead, and P. J. Headrick 1991. Engraved Cobbles from Early Archaeological Contexts in Central Texas. *Current Research in the Pleistocene* 8: 13–15.

COLLINS, M.B., and A. Kerr 1993. Archeology of the Earliest Texans. Paper presented at the annual meeting of the Texas Archeological Society, Laredo, Texas, 31 October 1993.

COTTERELL, B., and J. Kamminga 1987. The Formation of Flakes. *American Antiquity* 52: 675–708.

1990. *Mechanics of Pre-Industrial Technology.* Cambridge University Press, Cambridge.

COX, S.L. 1986. A Re-Analysis of the Shoop Site. *Archaeology of Eastern North America* 14: 101–170.

CRABTREE, D.E. 1968. Mesoamerican Polyhedral Cores and Prismatic Blades. *American Antiquity* 33: 446–478.

1972. *An Introduction to Flintworking*. Occasional Papers of the Idaho Museum of Natural History, No. 28. Pocatello.

1973. The Obtuse Angle as a Functional Edge. *Tebiwa* 17 (1): 1–6.

1982. *An Introduction to Flintworking*. 2d ed. Occasional Papers of the Idaho Museum of Natural History, No. 28. Pocatello.

CRABTREE, D.E., and E. H. Swanson 1968. Edge-Ground Cobbles and Blade-Making in the Northwest. *Tebiwa* 11 (2): 50–54.

CROOK, W.W., and R. K. Harris 1957. Hearths and Artifacts of Early Man near Lewisville, Texas, and Associated Faunal Material. *Bulletin of the Texas Archeological Society* 28: 7–97.

CURRAN, M.L. 1984. The Whipple Site and Paleoindian Tool Assemblage Variation: A Comparison of Intrasite Structuring. *Archaeology of Eastern North America* 12: 5–40.

DE HEINZELIN DE BRAUCOURT, J. 1962. *Manuel de typologie de industries lithiques*. L'Institut Royal des Sciences Naturelles de Belgique, Brussels.

DE SONNEVILLE-BORDES, D. 1960. *Le Paleolithique superieur en Perigord*. Delmas, Bordeaux.

1989. Foyers Paleolithiques en Perigord. In *Nature et fonction des foyers prehistoriques,* directed by M. Olive and Y. Taborin, pp. 225–237. Memoirs du Musee de Prehistoire d'Ille de France No. 2. Nemours.

DeJARNETTE, D.L., E. B. Kurjack, and J. W. Cambron 1962. Excavations at the Stanfield-Worley Bluff Shelter. *Journal of Alabama Archaeology* 8 (1–2): 1–124.

DELLER, D.B., and C. J. Ellis 1984. Crowfield: A Preliminary Report on a Probable Paleo-Indian Cremation in Southwestern Ohio. *Archaeology of Eastern North America* 12: 41–71.

1992. *Thedford II, A Paleo-Indian Site in the Ausable River Watershed of Southwestern Ontario*. Memoirs of the Museum of Anthropology, The University of Michigan 24. Ann Arbor.

DIBBLE, D.S., and D. Lorrain 1968. *Bonfire Shelter: A Stratified Bison Kill Site, Val Verde County, Texas*. Texas Memorial Museum, Miscellaneous Papers 1. The University of Texas at Austin.

DINCAUZE, D.F. 1993. Fluted Points in the Eastern Forests. In *From Kostenki to Clovis,* edited by O. Soffer and N. D. Praslov, pp. 279–292. Plenum Press, New York.

DIXON, E.J. 1993. *Quest for the Origins of the First Americans*. University of New Mexico Press, Albuquerque.

DRAGOO, D.W. 1973. Wells Creek—An Early Man Site in Stewart County, Tennessee. *Archaeology of Eastern North America* 1: 1–56.

DRISKELL, B.N. 1994. Stratigraphy and Chronology at Dust Cave. *Journal of Alabama Archaeology* 40: 17–33.

DUMONT, J. 1982. The Quantification of Microwear Traces: A New Use for Interferometry. *World Archaeology* 14: 206–217.

DUNBAR, J.S. 1991. Resource Orientation of Clovis and Suwannee Age Paleoindian Sites in Florida. In *Clovis: Origins and Adaptations,* edited by R. Bonnichsen and K. L. Turnmire, pp. 185–213. Center for the Study of the First Americans, Department of Anthropology, Oregon State University, Corvallis.

ELLIS, C.J., and D. B. Deller 1988. Some Distinctive Paleo-Indian Tool Types from the Lower Great Lakes Region. *Midcontinental Journal of Archaeology* 13: 111–158.

ELLIS, H.H., 1939. *Flint-Working Techniques of the American Indians: An Experimental Study.* Lithic Laboratory, Department of Anthropology, Ohio State University, Columbus.

EULER, R.C., and A. P. Olson 1965. Split-Twig Figurines from Northern Arizona: New Radiocarbon Dates. *Science* 148: 368–369.

EVANS, G.L. 1951. Prehistoric Wells in Eastern New Mexico. *American Antiquity* 17: 1–9.

FERRING, R. 1989. The Aubrey Clovis Site: A Paleo-Indian Locality in the Upper Trinity River Basin, Texas. *Current Research in the Pleistocene* 6: 9–11.

1990. The 1989 Investigations at the Aubrey Clovis Site, Texas. *Current Research in the Pleistocene* 7: 10–12.

1994. The Role of Geoarchaeology in Paleoindian Research. In *Method and Theory for Investigating the Peopling of the Americas,* edited by R. Bonnichsen and D. G. Steele, pp. 57–72. Center for the Study of the First Americans, Department of Anthropology, Oregon State University, Corvallis.

FISH, R.P. 1979. *The Interpretive Potential of Mousterian Debitage.* Arizona State University, Anthropological Research Papers 16. Tempe.

FLENNIKEN, J.J. 1985. Stone Tool Reduction Techniques as Cultural Markers. In *Stone Tool Analysis: Essays in Honor of Don E. Crabtree,* edited by M. G. Plew, J. C. Woods, and M. G. Pavesic, pp. 265–276. University of New Mexico Press, Albuquerque.

FORD, J.A. 1969. *A Comparison of Formative Cultures in the Americas: Diffusion or the Psychic Unity of Man.* Smithsonian Institution, Washington D.C.

FITTING, J.E., J. DeVisscher, and E. J. Wahla 1966. The Paleo-Indian Occupation of the Holcombe Beach. Museum of Anthropology, the University of Michigan, Anthropological Papers 27. Ann Arbor.

FOWLER, W. 1991. Lithic Analysis as a Means of Processual Inference in Southern Mesoamerica: A Review of Recent Research. In *Maya Stone Tools,* edited by T. R. Hester and H. J. Shafer, pp. 1–19. Monographs in World Archaeology 1. Prehistory Press, Madison.

FRISON, G.C. 1979. Observations on the Use of Stone Tools: Dulling of Working Edges of Some Chipped Stone Tools in Bison Butchering. In *Lithic Use-Wear Analysis,* edited by B. Hayden, pp. 259–268. Academic Press, New York.

1982. The Sheaman Site: A Clovis Component. In *The Agate Basin Site,* by G. C. Frison and D. J. Stanford, pp. 143–157. Academic Press, New York.

1986. Mammoth Hunting and Butchering from a Perspective of African Elephant Culling. In *The Colby Mammoth Site,* edited by G. C. Frison and L. C. Todd, pp. 115–134. University of New Mexico Press, Albuquerque.

1989. Experimental Use of Clovis Weaponry and Tools on African Elephants. *American Antiquity* 54: 766–784.

1991. The Clovis Cultural Complex: New Data from Caches of Flaked Stone and Worked Bone Artifacts. In *Raw Material Economies among Prehistoric Hunter-Gatherers,* edited by A. Montet-White and S. Holen, pp. 321–333. University of Kansas, Publications in Anthropology 19. Lawrence.

1993 North American High Plains Paleo-Indian Hunting Strategies and Weaponry Assemblages. In *From Kostenki to Clovis,* edited by O. Soffer and N. D. Praslov, pp. 237–249. Plenum Press, New York.

FRISON, G.C., and L. C. Todd 1986. *The Colby Mammoth Site: Taphonomy and Archaeology of a Clovis Kill in Northern Wyoming.* University of New Mexico Press, Albuquerque.

FUNK, R.E. 1977. Early Cultures in the Hudson Drainage Basin. In *Amerinds and Their Paleoenvironments in Northeastern North America,* edited by W. S. Newman and B. Salwen, pp. 316–332. Annals of the New York Academy of Sciences 288.

FUNK, R.E., and B. Wellman 1984. The Corditaipe Site: A Small Isolated Paleo-Indian Camp in the Upper Mohawk Valley. *Archaeology of Eastern North America* 12: 72–77.

GAXIOLA, M., and J. E. Clark (editors) 1989. *La obsidiana en Mesoamerica.* Serie Arqueologia, Instituto Nacional de Antropologia e Historia, Mexico City.

GOODE, G.T., and R. J. Mallouf 1991. The Evant Cores: Polyhedral Blade Cores from North-Central Texas. *Current Research in the Pleistocene* 8: 67–70.

GOULD, F.W. 1975. *Texas Plants: A Checklist and Ecological Summary.* Texas Agricultural Experiment Station, Texas A&M University, College Station.

GOULD, R.A. 1980. *Living Archaeology.* Cambridge University Press, Cambridge.

GRAHAM, R.W., C. V. Haynes, Jr., D. L. Johnson, and M. Kay 1981. Kimmswick: A Clovis-Mastodon Association in Eastern Missouri. *Science* 213 (4512): 1115–1117.

GRAMLY, R.M. 1982. *The Vail Site: A Palaeo-Indian Encampment in Maine.* Bulletin of the Buffalo Society of Natural Sciences 30. Buffalo.

1993. *The Richey Clovis Cache: Earliest Americans along the Columbia River.* Persimmon Press, Buffalo.

GRAMLY, R.M., and J. Lothrop 1984. Archaeological Investigations of the Potts Site, Oswego County, New York, 1982 and 1983. *Archaeology of Eastern North America* 12: 122–159.

GRAMLY, R.M., and K. Rutledge 1981. A New Paleo-Indian Site in the State of Maine. *American Antiquity* 46: 354–361.

GREEN, F.E. 1963. The Clovis Blades: An Important Addition to the Llano Complex. *American Antiquity* 29: 145–165.

GRIMES, J.R., W. Eldridge, B. G. Grimes, A. Vaccaro, F. Vaccaro, J. Vaccaro, N. Vaccaro, and A. Orsini 1984. Bull Brook II. *Archaeology of Eastern North America* 12: 159–183.

HAMMATT, H.H. 1969. Paleo-Indian Blades from Western Oklahoma. *Bulletin of the Texas Archeological Society* 40: 193–198.

1970. A Paleo-Indian Butchering Kit. *American Antiquity* 35: 141–152.

HAMMOND, N. 1976. Maya Obsidian Trade in Southern Belize. In *Maya Lithic Studies,* edited by T. R. Hester and N. Hammond, pp. 71–82. Center for Archaeological Research, The University of Texas at San Antonio, Special Report 4.

HART, A. 1983. Eleven Caches from near Lamesa, Texas. *Transactions of the 18th Regional Archaeological Symposium for Southeastern New Mexico and Western Texas,* pp. 1–150.

HAURY, E.W., E. B. Sayles and W. W. Wasley 1959. The Lehner Mammoth Site, Southeastern Arizona. *American Antiquity* 19: 1–14.

HAYNES, C.V., JR. 1966. Elephant-hunting in North America. *Scientific American* 214: 104–112.

1982. Were Clovis Progenitors in Beringia? In *Paleoecology of Beringia,* edited by D. M. Hopkins, J. V. Matthews, Jr., C. W. Schweger, and S. B. Young, pp. 383–398. Academic Press, New York.

1992. Contributions of Radiocarbon Dating to the Geochronology of the Peopling of the New World. In *Radiocarbon after Four Decades,* edited by R. E. Taylor, A. Long, and R. S. Kra, pp. 355–374. Springer-Verlag, New York.

1993. Clovis-Folsom Geochronology and Climatic Change. In *From Kostenki to Clovis,* edited by O. Soffer and N. D. Praslov, pp. 219–236. Plenum Press, New York.

HEMMINGS, T.E. 1970. *Early Man in the San Pedro Valley, Arizona.* Unpublished Ph.D. dissertation, University of Arizona, Tucson.

HENDERSON, J., and G. T. Goode 1991. Pavo Real: An Early Paleoindian Site in South-Central Texas. *Current Research in the Pleistocene* 8: 26–28.

HESTER, J.J. 1972. *Blackwater Locality No. 1: A Stratified Early Man Site in Eastern New Mexico.* Fort Burgwin Research Center Publication 8. Ranchos de Taos, New Mexico.

HESTER, T.R. 1972. Notes on Large Obsidian Blade Cores and Core-Blade Technology in Mesoamerica. *Contributions of the University of California Archaeological Research Facility* 14: 95–106.

1988. Notes on South Texas Archaeology: 1988–2. Studies of an Obsidian Clovis Point from the Central Texas Coast, and Other Paleo-Indian Obsidian Artifacts from Texas. *La Tierra* 15 (2): 2–3.

HESTER, T.R. (EDITOR) 1978. *Archaeological Studies of Mesoamerican Obsidian.* Ballena Press, Socorro, New Mexico.

HESTER, T.R., and D. M. Brown 1988. A Cache of Bifaces from Southern Texas. *La Tierra* 12 (4): 3–5.

HESTER, T.R., M. B. Collins, and P. J. Headrick 1992. Notes on South Texas Archaeology: 1992–4. Paleo-Indian Engraved Stones from the Gault Site. *La Tierra* 19 (4): 3–5.

HESTER, T.R., M. B. Collins, D. A. Story, E. S. Turner, P. Tanner, K. M. Brown, L. D. Banks, D. Stanford, and R. J. Long 1992. Paleoindian Archaeology at McFaddin Beach, Texas. *Current Research in the Pleistocene* 9: 20–22.

HESTER, T.R., G. L. Evans, F. Asaro, F. Stross, T. N. Campbell, and H. Michel 1985. Trace Element Analysis of an Obsidian Paleo-Indian Projectile Point from Kincaid Rockshelter, Texas. *Bulletin of the Texas Archeological Society* 56: 143–153.

HESTER, T.R., and N. Hammond 1976. Maya Lithic Studies: Papers from the 1976 Belize Field Symposium. Center for Archaeological Research, The University of Texas at San Antonio, Special Report 4.

HESTER, T.R., R. F. Heizer, and R. N. Jack 1971. Technology and Geological Sources of Obsidian from Cerro de las Mesas, Veracruz, Mexico, with Observations on Olmec Trade. *In Papers on Olmec and Maya Archaeology,* pp. 133–142. Contributions of the University of California Archaeological Research Facility 13. University of California at Berkeley.

HESTER, T.R., and H. J. Shafer 1975. An Initial Study of Blade Technology on the Central and Southern Texas Coast. *Plains Anthropologist* 20 (69): 175–185.

1987. Observations on Ancient Maya Core Technology and Colha, Belize. In *The Organization of Core Technology,* edited by J. K. Johnson and C. A. Morrow, pp. 239–258. Westview Press, Boulder.

HESTER, T.R., H. J. Shafer, J. D. Eaton, R. E. W. Adams, and G. Ligabue 1983. Colha's Stone Tool Industry. *Archaeology* 36: 45–52.

HESTER, T.R., H. J. Shafer, T. C. Kelly, and G. Ligabue 1982. Observations on the Patination Process and the Context of Antiquity: A Fluted Projectile Point from Belize, Central America. *Lithic Technology* 11:29–34.

HEIZER, R.F., and L. K. Napton 1970. Archaeological Investigations in Lovelock Cave, Nevada. In *Archaeology and the Prehistoric Great Basin Lacustrine Subsistence Regime as Seen from Lovelock Cave, Nevada,* edited by R. F. Heizer and L. K. Napton, pp. 1–86. Contributions of the University of California Archaeological Research Facility 10. Berkeley.

HOFFMAN, R., and L. Gross 1970. Reflected-Light Differential-Interference Microscopy: Principles, Use and Image Interpretation. *Journal of Microscopy* 91 (3): 149–172.

HOFMAN, J.L., R. L. Brooks, J. S. Hays, D. W. Owsley, R. L. Jantz, M. K. Marks, and M. H. Manhein 1989. *From Clovis to Comanchero: Archeological Overview of the Southern Great Plains.* Arkansas Archeological Survey Research Series 35. Fayetteville.

HOLLIDAY, V.T. 1992. Soil Formation, Time, and Archaeology. In *Soils in Archaeology,* edited by V. T. Holliday, pp. 101–117. Smithsonian Institution Press, Washington, D.C.

1997. *Paleoindian Geoarchaeology of the Southern High Plains.* University of Texas Press, Austin.

HOLLIDAY, V.T., C. V. Haynes, Jr., J. L. Hofman, and D. J. Meltzer 1994. Geoarchaeology and Geochronology of the Miami (Clovis) Site, Southern High Plains of Texas. *Quaternary Research* 41: 234–244.

HUBBERT, C.M. N.D. Untitled manuscript prepared for the Tennessee Valley Authority on cultural resources in Pickwick Basin, Alabama, 1980. Manuscript in author's files.

1989. Paleo-Indian Settlement in the Middle Tennessee Valley: Ruminations from the Quad Paleo-Indian Locale. *Tennessee Anthropologist* 14 (2): 148–164.

HUCKELL, B.B. N.D. Clovis Lithic Technology: A View from the Upper San Pedro Valley. Manuscript in possession of the author.

HUMPHREY, J.D., and C. R. Ferring 1994. Stable Isotopic Evidence for Latest Pleistocene and Holocene Climatic Change in North-Central Texas. *Quaternary Research* 41: 200–213.

IDELL, A. 1957. *The Bernal Diaz Chronicles.* Doubleday, Garden City N.Y.

INIZAN, M.-L., H. Roche, and J. Tixier. 1992. Technology of Knapped Stone. *Préhistoire de la pierre taillée,* Tome 3. Cercle de Recherches et d'Etudes Prehistorique avec le concours du Centre National de la Recherche Scientifique. Meudon.

IRWIN, H.T. 1971. Developments in Early Man Studies in Western North America, 1960–1970. *Arctic Anthropology* 8: 42–67.

JANES, S.M. 1930. Seven Trips to Mount Livermore. *West Texas Historical and Scientific Society Publications* 3: 8–9. Sul Ross State Teachers College. Alpine.

JETT, S.C. 1968. Grand Canyon Dams, Split-Twig Figurines, and "Hit-and-Run" Archaeology. *American Antiquity* 33: 341–351.

JOHNSON, E. 1987. *Lubbock Lake: Late Quaternary Studies on the Southern High Plains.* Texas A&M University Press, College Station. 1991. Late Pleistocene Cultural Occupation on the Southern Plains. In *Clovis Origins and Adaptations.* edited by R. Bonnichsen and K. L. Turnmire, pp. 215–236. Center for the Study of the First Americans, Department of Anthropology, Oregon State University, Corvallis.

JOHNSON, E., and V. T. Holliday 1984. Comments on "Large Springs and Early American Indians" by Joel Shiner. *Plains Anthropologist* 29: 65–70.

JOHNSON, L., JR. 1991. *Early Archaic Life at the Sleeper Archaeological Site, 41BC65, of the Texas Hill Country, Blanco County, Texas.* Texas State Department of Highways and Public Transportation, Publications in Archaeology, Report 39. Austin.

JUDGE, W.J. 1973. *Paleoindian Occupation of the Central Rio Grande Valley in New Mexico.* University of New Mexico Press, Albuquerque.

JULIG, P.J. 1984. Cummins Paleo-Indian Site and Its Paleo-environments, Thunder Bay, Ontario. *Archaeology of Eastern North America* 12: 192–209.

KAY, M. 1996. Microwear Analysis of Some Clovis and Experimental Chipped Stone Tools. In *Stone Tools: Theoretical Insights into Human Prehistory,* edited by G. C. Odell, pp. 315–344. Plenum, New York. 1997. Imprints of Ancient Tool Use at Monte Verde, In *Monte Verde: A Late Pleistocene Settlement in Chile,* by T. D. Dillehay, pp. 649–660. Smithsonian Press, Washington, D.C. N.D.. Blackwater Draw Mitchell Locality (LA3324) Microwear Evaluations. Manuscript on file, Office of Conservation Archaeology, University of Pittsburgh at Greensburg, Pennsylvania.

KEELEY, L.H. 1980. *Experimental Determination of Stone Tool Use: A Microwear Analysis.* University of Chicago Press. Chicago.

KEENLYSIDE, D.L. 1991. Paleoindian Occupations of the Maritimes Region of Canada. In *Clovis: Origins and Adaptations,* edited by R. Bonnichsen and K. L. Turnmire, pp. 163–173. Center for the Study of the First Americans, Department of Anthropology, Oregon State University, Corvallis.

KELLY, R.L., and L. C. Todd 1988. Coming into the Country: Early Paleoindian Hunting and Mobility. *American Antiquity* 53: 231–244.

KELLY, T.C. 1992. Two Polyhedral Cores from Comanche Hill, San Antonio, Texas. *La Tierra* 19 (2): 29–33.

KLIPPEL, W.E. 1971. *Graham Cave Revisited.* Missouri Archaeological Society, Memoir 9.

KOOB, S.P. 1986. The Use of Palaloid B-72 as an Adhesive: Its Application for Archaeological Ceramics and Other Materials. *Studies in Conservation* 31: 7–14.

KORNFELD, M., K. Akoshima, and G. C. Frison 1990. Stone Tool Caching on the North American Plains: Implications of the McKean Site Tool Kit. *Journal of Field Archaeology* 17: 301–309.

KRAFT, H.C. 1973. The Plenge Site: A Paleo-Indian Occupation in New Jersey. *Archaeology of Eastern North America* 1: 56–117.

KRIEGER, A.D. 1947. Certain Projectile Points of the Early American Hunters. *Bulletin of the Texas Archeological and Paleontological Society* 18: 7–27.

KUCHLER, A.W. 1964. *Potential Natural Vegetation of the Coterminous United States.* American Geographical Society, Special Publication 36.

LAHREN, L., and R. Bonnichsen 1974. Bone Foreshafts from a Clovis Burial in Southwestern Montana. *Science* 186: 147–150.

LEE, R.B. 1979. *The !Kung San: Men, Women, and Work in a Foraging Society.* Cambridge University Press, Cambridge.

LEONHARDY, F.C. 1966. *Domebo: A Paleo-Indian Mammoth Kill in the Prairie-Plains.* Contributions of the Museum of the Great Plains 1. Lawton.

LEPPER, B.T., and D. J. Meltzer 1991. Late Pleistocene Human Occupation of the Eastern United States. In *Clovis: Origins and Adaptations,* edited by R. Bonnichsen and K. L. Turnmire, pp. 176–184. Center for the Study of the First Americans, Department of Anthropology, Oregon State University, Corvallis.

LEROI-GOURHAN, A. 1943. Evolution et techniques. *I: L'homme et la matiere.* A. Michel, Paris.

LOGAN, W.D. 1952. *Graham Cave.* Missouri Archaeological Society, Memoir 2.

LONG, R.J. 1977. *McFaddin Beach.* The Patillo Higgins Series of Natural History and Anthropology 1. Spindletop Museum, Lamar University, Beaumont.

LOWERY, D. 1989. The Paw Paw Cove Paleoindian Site Complex, Talbot County, Maryland. *Archaeology of Eastern North America* 17: 143–163.

LUEDTKE, B. 1992. *An Archaeologist's Guide to Chert and Flint.* University of California Press, Los Angeles.

McCARY, B.C. 1951. A Workshop Site of Early Man in Dinwiddie County, Virginia. *American Antiquity* 17: 9–17.

McCORMICK, O. 1982. 41WM419 (Crockett Gardens Site). In *Archaeological Investigations at the San Gabriel Reservoir Districts, Central Texas,* compiled and edited by T. R. Hays. pp. 12.135–12.166. Archaeology Program, Institute of Applied Sciences, North Texas State University, Denton.

McDONALD, G.F. 1966. The Technology and Settlement Pattern of a Paleo-Indian Site at Debert, Nova Scotia. *Quaternaria* 8: 59–74.

1968. *Debert, A Paleo-Indian Site in Central Nova Scotia.* National Museums of Canada, Anthropology Papers 16. Ottawa.

1971. A Review of Research on Paleo-Indian in Eastern North America, 1960–1970. *Arctic Anthropology* 8 (2): 32–41.

McGREGOR, D.E. 1987. Lithic Raw Material Utilization. In *Hunter-Gatherer Adaptations along the Prairie Margin: Site Excavations and Synthesis of Prehistoric Archaeology, pp. 186–195.* Richland Creek Technical Series, Vol. 3. Archaeology Research Program, Institute for the Study of Earth and Man, Southern Methodist University, Dallas.

McNETT, C.W., JR. 1985. Artifact Morphology and Chronology at the Shawnee Minisink Site. In *Shawnee Minisink: A Stratified Paleoindian—Archaic Site in the Upper Delaware Valley of Pennsylvania,* edited by C. W. McNett, Jr., pp. 83–120. Academic Press, Orlando.

McNETT, C.W. JR., B. A. McMillan, and S. B. Marshall 1977. The Shawnee-Minisink Site. In *Amerinds and their Paleoenvironments in Northeastern North America,* edited by W. S. Newman and B. Salwen, pp. 282–296. Annals of the New York Academy of Sciences 288.

McSWAIN, R. 1991. A Comparative Evaluation of the Producer-Consumer Model for Lithic Exchange in Northern Belize, Central America. *Latin American Antiquity* 2: 337–351.

MALLOUF, R.J. 1981. *A Case Study of Plow Damage to Chert Artifacts: The Brookeen Creek Cache, Hill County, Texas.* Office of the State Archeologist, Report 33. Texas Historical Commission, Austin.

1982. An Analysis of Plow-damaged Chert Artifacts: The Brookeen Creek Cache (41HI86), Hill County, Texas. *Journal of Field Archaeology* 9 (1): 79–98.

1989. A Clovis Quarry Workshop in the Callahan Divide: The Yellow Hawk Site, Taylor County, Texas. *Plains Anthropologist* 34 (124): 81–103.

1994. Sailor-Helton: A Paleoindian Cache from Southwestern Kansas. *Current Research in the Pleistocene* 11: 44–46.

MALLOUF, R.J., and B. J. Baskin 1976. *Archeological Surveys in the Tehuacana Creek Watershed, Hill and McLennan Counties, Texas.* Office of the State Archeologist, Survey Report 19. Texas Historical Commission, Austin.

MALLOUF, R.J., and V. A. Wulfkuhle 1991. Notes on the Helton-Harrel Biface Cache from Seward County, Kansas. *Kansas Anthropologist* 12 (1): 1–12.

MASON, R. 1962. The Paleo-Indian Tradition in Eastern North America. *Current Anthropology* 3: 227–283.

MASSON, M.A., and M. B. Collins 1995. The Wilson-Leonard Site (41WM235). *Cultural Resource Management News and Views* 7 (1): 6–10.

MEADE, D.W., W. G. Chervenda, and J. M. Greenwade 1974. *Soil Survey of Navarro County, Texas.* U.S.D.A. Soil Conservation Service, Washington, D.C.

MEEKS, S.C. 1994. Lithic Artifacts from Dust Cave. *Journal of Alabama Archaeology* 40: 77–103.

MEHRINGER, P.J. 1988. Weapons Cache of Ancient Americans. *National Geographic* 174: 500–503.

1989. Of Apples and Archaeology. *Universe* 1 (2): 2–8.

MEHRINGER, P.J., and F. F. Foit, Jr. 1990. Volcanic Ash Dating of the Clovis Cache at East Wenatchee, Washington. *National Geographic Research* 6: 495–503.

MELTZER, D.J. 1986. A Study of Texas Clovis Points. *Current Research in the Pleistocene* 3: 33–36.

1987. The Clovis Paleoindian Occupation of Texas: Results of the Texas Clovis Fluted Point Survey. *Bulletin of the Texas Archeological Society* 57: 27–68.

1989. An Update on the Texas Clovis Fluted Point Survey. *Current Research in the Pleistocene* 6: 31–34.

1989. Was Stone Exchanged among Eastern North American Paleo-Indians? In *Eastern Paleo-Indian Lithic Resource Use,* edited by C. Ellis and J. Lothrop, pp. 11–39. Westview Press, Boulder.

1993. Is There a Clovis Adaptation? In *From Kostenki to Clovis* edited by O. Soffer and N. D. Praslov, pp. 293–310. Plenum Press, New York.

MILLER, K.A. 1993. *A Study of Prehistoric Biface Caches from Texas.* Unpublished master's thesis, The University of Texas at Austin.

MONTGOMERY, J., and J. Dickenson 1992. Additional Blades from Blackwater Draw Locality No. 1, Portales, New Mexico. *Current Research in the Pleistocene* 9: 32–33.

MORROW, J.E. 1995. Clovis Projectile Point Manufacture: A Perspective from the Ready/Lincoln Hills Site, 11JY46, Jersey County, Illinois. *Midcontinental Journal of Archaeology* 20 (2): 167–191.

MORSE, D.F., and P. A. Morse 1983. *Archaeology of the Central Mississippi Valley.* Academic Press, New York.

MOVIUS, H.L., JR. 1968. The Hearths of the Upper Perigordian and Aurignacian Horizons at the Abri Pataud, Les Eyzies (Dordogne), and Their Possible Significance. In *Recent Studies in Paleoanthropology,* edited by J. D. Clark and F. C. Howell, pp. 296–325. American Anthropologist 68 (2) Part 2.

NELSON, F.W., and J. E. Clark 1990. The Determination of Exchange Patterns in Prehistoric Mesoamerica. In *Nuevos Enfoques en el Estudio de la Litica,* edited by M. de los Dolores Soto de Arechavaleta, pp. 153–175. Universidad Nacional Autonoma de Mexico. Mexico City.

NELSON, R.W., K. K. Nielson, N. F. Mangelson, M. W. Hill, and R. T. Matheny 1977. Preliminary Studies of the Trace Element Composition of Obsidian Artifacts from Northern Campeche, Mexico. *American Antiquity* 42: 209–225.

NEVES, W.A., and H. M. Pucciarelli 1991. Morphological Affinities of the First Americans: An Exploratory Analysis Based on Early South American Human Remains. *Journal of Human Evolution* 21: 261–273.

NEWCOMER, M.H. 1971. Some Quantitative Experiments in Handaxe Manufacture. *World Archaeology* 3: 85–93.
1975. "Punch Technique" and Upper Palaeolithic Blades. In *Lithic Technology: Making and Using Stone Tools,* edited by E. Swanson, pp. 97–102. Mouton, The Hague.

NIETO CALLEJA, R., and F. Lopez Aguillar 1990. Los contextos arqueologicos en yacimientos de obsidiana. In *Nuevos enfoques en el estudio de la litica,* edited by M. de los Dolores Soto de Arechavaleta, pp. 177–214. Universidad Nacional Autonoma de Mexico, Mexico City.

OWEN, L.R. 1988. *Blade and Microblade Technology.* BAR International Series 441. Oxford.

PASTRANO, A. 1990. Produccion de instrumentos en obsidiana. Division del trabajo (Proyecto Tula). In *Nuevos enfoques en el estudio de la litica,* edited by M. de los Dolores Soto de Arechavaleta, pp. 243–296. Universidad Autonoma de Mexico, Mexico City.

PERLES, C. 1976. Le Feu. In *La prehistoire Française,* directed by H. de Lumley, pp. 679–683. Editions du Centre National de la Recherche Scientifique, Paris.

PHILLIPS, P. 1988. Traceology (Microwear) Studies in the USSR. *World Archaeology* 19: 349–356.

PIRES-FERREIRA, J. 1976. Obsidian Exchange in Formative Mesoamerica. In *The Early Mesoamerican Village,* edited by K. V. Flannery, pp. 292–306. Academic Press, New York.

PRESTON, D. 1997. The Lost Man. *New Yorker* June 16, 1997: 70–81.

PREWITT, E.R. 1981. A Wooden Mortar from the Stockton Plateau of Texas. *Journal of Field Archaeology* 8: 111–117.

PURDY, B.A. 1981. *Florida's Prehistoric Stone Technology.* University Presses of Florida, Gainesville.

RANERE, A.J., and R. G. Cooke 1991. Paleoindian Occupation in the Central American Tropics. In *Clovis: Origins and Adaptations,* edited by R. Bonnichsen and K. L. Turnmire, pp. 237–253. Center for the Study of the First Americans, Department of Anthropology, Oregon State University, Corvallis.

RAY, C.N., and K. Bryan 1938. Folsomoid Point Found in Alluvium beside a Mammoth's Bones. *Science* 88: 257–276.

REDDER, A.J. 1985. Horn Shelter Number 2: The South End, A Preliminary Report. *Central Texas Archeologist* 10: 37–65.

RICKLIS, R.A. 1995. Toyah Components: Evidence for Occupation in the Project Area During the Latter Part of the Late Prehistoric Period. Chapter 8 in *Archaic and Late Prehistoric Human Ecology in the Middle Onion Creek Valley, Hays County, Texas,* by R. A. Ricklis and M. B. Collins. Texas Archeological Research Laboratory, Studies in Archeology 19. The University of Texas at Austin.

RICKLIS, R.A., M. D. Blum, and M. B. Collins 1991. *Archeological Testing at the Vera Daniel Site (41TV1364) Zilker Park, Austin, Texas.* Texas Archeological Research Laboratory, Studies in Archeology 12. The University of Texas at Austin.

RITCHIE, W.A. 1953. A Probable Paleo-Indian site in Vermont. *American Antiquity* 18: 249–258.

1965. *The Archaeology of New York State.* Natural History Press, New York.

ROOSA, W.B. 1977. Great Lakes Paleoindian: The Parkhill Site, Ontario. In *Amerinds and Their Paleoenvironments in Northeastern North America,* edited by W. S. Newman and B. Salwen, pp. 349–354. Annals of the New York Academy of Sciences 288.

SANDERS, T.N. 1990. *Adams: The Manufacturing of Flaked Stone Tools at a Paleoindian Site in Western Kentucky.* Persimmon Press, Buffalo.

SAUNDERS, J.L. 1992. Blackwater Draws: Mammoths and Mammoth Hunters in the Terminal Pleistocene. In *Proboscidean and Paleoindian Interactions,* edited by J. W. Fox, C. B. Smith, and K. T. Wilkins, pp. 123–147. Baylor University Press, Waco.

SCHICK, K.D., and N. Toth 1993. *Making Silent Stones Speak.* Simon and Schuster, New York.

SCHIFFER, M.B. 1987. *Formation Processes of the Archaeological Record.* University of New Mexico Press, Albuquerque.

SCHLANGER, S.H. 1981. Tool Caching Behavior and the Archaeological Record. Paper presented at 46th Annual Meeting of the Society for American Archaeology, San Diego.

SCHWARTZ, D.W., A. L. Lange, and R. deSaussure 1958. Split-Twig Figurines in the Grand Canyon. *American Antiquity* 23: 204–274.

SELLARDS, E.H. 1938. Artifacts Associated with Fossil Elephant. *Bulletin of the Geological Society of America* 49: 999–1010.

1952. *Early Man in America.* University of Texas Press, Austin.

SEMENOV, S.A. 1964. *Prehistoric Technology.* Cory, Adams and MacKay, London.

SHAFER, H.J. 1985. A Technological Study of Two Maya Lithic Workshops at Colha, Belize. In *Stone Tool Analysis: Essays in Honor of Don E. Crabtree,* edited by M. Plew, J. Woods, and M. Pavesic, pp. 277–315. University of New Mexico Press, Albuquerque.

1991. Late Preclassic Formal Stone Tool Production at Colha, Belize. In *Maya Stone Tools,* edited by T. R. Hester and H. J. Shafer, pp. 31–44. Prehistory Press, Madison.

SHAFER, H.J., and T. R. Hester 1983. Ancient Maya Chert Workshops in Northern Belize, Central America. *American Antiquity* 48: 519–543.

1986. Maya Stone-Tool Craft Specialization and Production at the Maya site of Colha, Belize: Reply to Mallory. *American Antiquity* 51: 158–166.

1991. Lithic Craft Specialization and Product Distribution at the Maya site of Colha Belize. *World Archaeology* 23: 79–97.

SHAWCROSS, W. 1976. Kauri Point Swamp: The Ethnographic Interpretation of a Prehistoric Site. In *Problems in Economic and Social Archaeology,* edited by G. de G. Sieveking, I. H. Longworth, and K. E. Wilson, pp. 277–305. Duckworth, London.

SHEETS, P.D. 1972. A Model of Mesoamerican Obsidian Technology Based on Preclassic Workshop Debris in El Salvador. *Ceramica de Cultura Maya* 8: 17–33.

SHEETS, P.D., K. Firth, F. Lange, F. Stross, F. Asaro, and H. Michel 1990. Obsidian Sources and Elemental Analyses of Artifacts in Southern Mesoamerica and the Northern Intermediate Area. *American Antiquity* 55: 144–158.

SHEETS, P.D., and G. R. Muto 1972. Pressure Blades and Total Cutting Edge: An Experiment in Lithic Technology. *Science* 175: 632–634.

SHINER, J.L. 1981. History, Economy, and Magic at a Freshwater Spring. In *In The Realms of Gold,* edited by W. A. Cockrell, pp. 202–203. Fathom Eight Special Publication 1. San Marino, California.

1982. Excavations at Aquarena Springs. In *Underwater Archaeology: The Proceedings of the Eleventh Conference on Underwater Archaeology,* edited by C. R. Cummings, p. 20. Fathom Eight Special Publication 4. San Marino, California.

1983. Large Springs and Early American Indians. *Plains Anthropologist* 28: 1–7.

SHIPMAN, P. 1992. Body Size and Broken Bones: Preliminary Interpretations of Proboscidean Remains. In *Proboscidean and Paleoindian Interactions,* edited by J. W. Fox, C. B. Smith, and K. T. Wilkins, pp. 75–98. Baylor University Press, Waco.

SHOTT, M.J. 1993. *The Leavitt Site: A Parkhill Phase Paleo-Indian Occupation in Central Michigan.* Museum of Anthropology, University of Michigan, Memoir 25. Ann Arbor.

SIMMONS, D.B., M. J. Shott, and H. T. Wright 1984. The Gainey Site: Variability in a Great Lakes Paleo-Indian Assemblage. *Archaeology of Eastern North America* 12: 266–279.

SLESICK, L.M. 1978. A Lithic Tool Cache in the Texas Panhandle. *Bulletin of the Texas Archeological Society* 49: 419–530.

SOFFER, O. 1985. *The Upper Paleolithic of the Central Russian Plain.* Academic Press, Orlando.

SOLLBERGER, J., and L. W. Patterson 1976. Prismatic Blade Replication. *American Antiquity* 41: 518–531.

SOTO DE ARECHAVALETA, M. DE LOS DOLORES 1990. Areas de actividad en un taller de manufactura de implementos de piedra tallada. In *Nuevos enfoques en el estudio de la litica,* edited by M. de los Dolores Soto de Arechavaleta, pp. 214–242. Universidad Nacional Autonoma de Mexico, Mexico City.

SPENCE, M.W. 1967. The Obsidian Industry of Teotihuacan. *American Antiquity* 32: 507–514.

1981. Obsidian Production and the State in Teotihuacan. *American Antiquity* 46: 769–788.

1990. El estado de investigaciones liticas en Mesoamerica. In *Nuevos enfoques en el estudio de la litica,* edited by M. de los Dolores Soto de Arechavaleta, pp. 431–442. Universidad Nacional Autonoma de Mexico, Mexico City.

SPETH, J. 1972. Mechanical Basis for Mechanical Flaking. *American Antiquity* 37: 34–60.

SPIESS, A.E., and D. B. Wilson 1987. *Michaud: A Paleoindian Site in the New England—Maritimes Region.* Occasional Papers in Maine Archaeology 6. Maine Historic Preservation Commission, Augusta.

STANFORD, D. 1982. A Critical Review of Archeological Evidence Relating to the Antiquity of Human Occupation of the New World. In *Plains Indian Studies: A Collection of Essays in Honor of John C. Ewers and Waldo R. Wedel,* edited by D. H. Ubelaker and H. J. Viola, pp. 202–218. Smithsonian Contributions to Anthropology 30. Smithsonian Press, Washington, D.C.

1983. Pre-Clovis Occupation South of the Ice Sheets. In *Early Man in the New World,* edited by R. Shutler, Jr., pp. 65–72. Sage Publications, Beverly Hills.

1991. Clovis Origins and Adaptations: An Introductory Perspective. In *Clovis: Origins and Adaptations,* edited by R. Bonnichsen and K. L. Turnmire, pp. 1–13. Center for the Study of the First Americans, Department of Anthropology, Oregon State University, Corvallis.

STANFORD, D.J. and M. A. Jodry 1988. The Drake Clovis Cache. *Current Research in the Pleistocene* 5: 21–22.

STARK, B.L., L. Heller, M. D. Glascock, J. M. Elam, and H. Neff 1992. Obsidian-Artifact Source Analysis for the Mixtequilla Region, South Central Veracruz, Mexico. *Latin American Antiquity* 3: 221–239.

STEELE, D.G., and D. L. Carlson 1989. Excavation and Taphonomy of Mammoth Remains from the Duewall-Newberry Site, Brazos County, Texas. In *Bone Modification,* edited by R. Bonnichsen and M. H. Sorg, pp. 413–430. Center for the Study of the First Americans, Institute for Quaternary Studies, University of Maine, Orono.

STEELE, D.G., and J. F. Powell 1994. Paleobiological Evidence of the Peopling of the Americas. In *Method and Theory for Investigating the Peopling of the Americas,* edited by R. Bonnichsen and D. G. Steele, pp. 141–164. Center for the Study of the First Americans, Department of Anthropology, Oregon State University, Corvallis.

STOLTMAN, J.B. 1971. Prismatic Blades from Northern Minnesota. *Plains Anthropologist* 16 (52): 105–109.

STORCK, P.L. 1983. The Fisher Site: Fluting Techniques and Early Palaeo-Indian Cultural Relationships. *Archaeology of Eastern North America* 11: 80–97.

1991. Imperialists without a State: The Cultural Dynamics of Early Paleoindian Colonization as Seen from the Great Lakes Region. In *Clovis: Origins and Adaptations,* edited by R. Bonnichsen and K. L. Turnmire, pp. 153–162. Center for the Study of the First Americans, Department of Anthropology, Oregon State University, Corvallis.

STORY. D.A. 1990. Cultural History of the Native Americans. In *The Archeology and Bioarcheology of the Gulf Coastal Plain,* Vol. 1., by D. A. Story, J. A. Guy, B. A. Burnett, M. D. Freeman, J. C. Rose, D. G. Steele, B. W. Olive, and K. J. Reinhard, pp. 163–366. Arkansas Archeological Survey, Research Series 38. Fayetteville.

TAKAC, P.R. 1991. Underwater Excavations at Spring Lake: A Paleoindian Site in Hays County, Texas. *Current Research in the Pleistocene* 8: 46–48.

TANKERSLEY, K.B. 1995. Paleoindian Contexts and Artifact Distribution Patterns at the Gostrom Site, St. Clair County, Illinois. *Midcontinental Journal of Archaeology* 20 (1): 40–61.

TIMMINS, P.A. 1994. Alder Creek: A Paleo-Indian Crowfield Phase Manifestation in the Region of Waterloo, Ontario. *Midcontinental Journal of Archaeology* 19: 170–197.

TIXIER, J. 1963. *Typologie de l'Epipaleolithique du Maghreb.* Memoires de Centre de Recherches Anthropologiques, Prehistoriques et Ethnographiques 2. Alger, Paris.

TOBEY, M.H. 1986. *Trace Element Analysis of Maya Chert from Belize.* Papers of the Colha Project 1. Center for Archaeological Research, The University of Texas at San Antonio.

TUNNELL, C. 1978. *The Gibson Lithic Cache from West Texas.* Office of the State Archeologist Report 30. Texas Historical Commission, Austin.

1989. Versatility of a Late Prehistoric Flint Knapper: The Weaver-Ramage Chert Cache of the Texas Rolling Plains. In *In the Light of Past Experience: Papers in Honor of Jack T. Hughes,* edited by B. C. Roper, pp. 369–397. Panhandle Archeological Society Publication 5. Canyon.

TURNER, E.S., and P. Tanner 1994. The McFaddin Beach Site on the Upper Texas Coast. *Bulletin of the Texas Archeological Society* 65: 319–336.

WATT, F.H. 1978. Radiocarbon Chronology of Sites in the Central Brazos Valley. *Bulletin of the Texas Archeological Society* 49: 111–138.

WHEELER, S.M. 1942. *Archeology of Etna Cave, Lincoln County, Nevada.* Nevada State Park Commission, Carson City.

WHITTAKER, J.C. 1994. *Flintknapping: Making and Understanding Stone Tools.* University of Texas Press, Austin.

WHITTHOFT, J. 1952. A Paleo-Indian Site in Eastern Pennsylvania: An Early Hunting Culture. *Proceedings of the American Philosophical Society* 96 (4): 464–495.

WILKE, P.J., J. J. Flenniken, and T. L. Ozbun 1991. Clovis Technology at the Anzick Site, Montana. *Journal of California and Great Basin Anthropology* 13 (2): 242–272.

WILKE, P.J., and L.A. Quintero 1994. Naviform Core-and-Blade Technology: Assemblage Character as Determined by Replicative Experiments. In *Neolithic Chipped Stone Industries of the Fertile Crescent,* edited by H. G. Gebel and S. K. Kozlowski, pp. 33–60. Studies in Near Eastern Production, Subsistence, and Environment 1. Ex oriente, Berlin.

WILLEY, G.R. 1966. *An Introduction to American Archaeology, Vol. 1: North and Middle America.* Prentice-Hall, Englewood Cliffs.

WILLEY, G.R., and P. Phillips 1958. *Method and Theory in American Archaeology.* University of Chicago Press, Chicago.

WILLEY, G.R., and J. A. Sabloff 1980. *A History of American Archaeology.* W. H. Freeman, San Francisco.

WILLIG, J.A. 1988. Paleo-Archaic Adaptations and Lakeside Settlement Patterns in the Northern Alkali Lake Basin, Oregon. In *Early Human Occupation in Far Western North America: The Clovis-Archaic Interface,* edited by J. A. Willig, C. M. Aikens, and J. L. Fagan, pp. 417–482. Nevada State Museum Anthropological Papers 21. Carson City.

1991. Clovis Technology and Adaptation in Far Western North America: Regional Pattern and Environmental Context. In *Clovis: Origins and Adaptations,* edited by R. Bonnichsen and K. L. Turnmire, pp. 91–118. Center for the Study of the First Americans, Department of Anthropology, Oregon State University, Corvallis.

WITTE, A.H. 1942. Certain Caches of Flints from the North Texas Area. *Bulletin of the Texas Archeological and Paleontological Society* 14: 72–76.

WOODS, J.C., and G. L. Titmus 1985. A Review of the Simon Clovis Collection. *Idaho Archaeologist* 8 (1): 3–8.

WORMINGTON, H.M. 1957. *Ancient Man in North America,* 4th ed. Denver Museum of Natural History, Denver.

WRIGHT, H.T., and W. B. Roosa 1966. The Barnes Site: A Fluted Point Assemblage from the Great Lakes Region. *American Antiquity* 31: 850–860.

YOUNG, B. 1988. Site Survey Form for Site 41NV659. On file, Texas Archeological Research Laboratory, The University of Texas at Austin.

YOUNG, B., and M. B. Collins 1989. A Cache of Blades with Clovis Affinities from Northeastern Texas. *Current Research in the Pleistocene* 6: 26–28.

Printed and bound by CPI Group (UK) Ltd, Croydon, CR0 4YY

27/10/2024

14580155-0001